Mike Yajko

Edward W. Orr

Performance Enhancement in Coatings

Hanser Publishers, Munich

Hanser/Gardner Publications, Inc., Cincinnati

The Author:
Edward W. Orr, BYK-Chemie USA, Wallingford, CT 06492, USA

Distributed in the USA and in Canada by
Hanser/Gardner Publications, Inc.
6915 Valley Avenue, Cincinnati, Ohio 45244-3029, USA
Fax: (513) 527-8950
Phone: (513) 527-8977 or 1-800-950-8977
Internet: http://www.hansergardner.com

Distributed in all other countries by
Carl Hanser Verlag
Postfach 86 04 20, 81631 München, Germany
Fax: +49 (89) 98 12 64
Internet: http://www.hanser.de

The use of general descriptive names, trademarks, etc., in this publication, even if the former are not especially identified, is not to be taken as a sign that such names, as understood by the Trade Marks and Merchandise Marks Act, may accordingly be used freely by anyone.

While the advice and information in this book are believed to be true and accurate at the date of going to press, neither the authors nor the editors nor the publisher can accept any legal responsibility for any errors or omissions that may be made. The publisher makes no warranty, express or implied, with respect to the material contained herein.

Library of Congress Cataloging-in-Publication Data
Orr, Edward W.
Additives for environmentally friendly coatings/Edward W. Orr.
  p. cm.
Includes index.
ISBN 1-56990-263-1
1. Coatings–Additives. I. Title.
TP156.C57077 1998
667′.9–dc21   98-30429

Die Deutsche Bibliothek – CIP-Einheitsaufnahme
Orr, Edward W.:
Performance enhancement in coatings/Edward W. Orr – Munich:
Hanser; Cincinnati: Hanser/Gardner, 1998
  ISBN 3-446-19405-3

© Carl Hanser Verlag, Munich 1998
Typeset in England by Alden Bookset, Oxford
Printed and bound in Germany by Kösel, Kempten

Dedicated to Dr. Shelby Freland Thames,
Distinguished University Research Professor, and to the
pioneering work performed by his organization at the
Shelby Freland Thames Polymer Science Research Center.

Preface

Over the last fifteen years, I have taught courses and held seminars on surfactants, polymers, and additives for environmentally friendly coatings and allied products. I have searched in vain for a textbook, but to no avail. There is absolutely no shortage of general coatings textbooks and raw material lists dealing with the rather eclectic aspects of pigments, resins, solvents, and additives—but I felt that a more concerted approach to the study of performance improvement in environmentally friendly coatings was necessary. After all, such coatings have evolved much more rapidly than ever thought possible, and they have consequently risen to clear dominance in the industry; nevertheless, they remain undoubtedly the most difficult to formulate and the least understood of all systems. Why is this? What makes environmentally friendly coatings so enigmatic, unique, and fascinating? And, perhaps the most important question of all—why is there such a shortage of integrated information?

These questions can be answered quite simply: the science of environmentally friendly coatings is a very young discipline. As a result of the rapid evolution that always accompanies science in its infancy, there is a quite unsurprising shortage of integrative information that overviews the interrelationships that exist among all the disparate subdisciplines. This book was designed to fill this gap, thus providing the missing link by offering a critical overview of cutting edge, interdisciplinary technology designed to improve performance. Furthermore, such technology is most effectively and efficiently described within the special context of additives—high performance ingredients expressly designed to interact with and enhance the performance of virtually all ingredients in the coating system—whether such ingredients be pigments, resins, solvents, or otherwise.

Of all the challenges encountered in the coatings industry, performance improvement—particularly in environmentally friendly coatings—is consistently ranked as the number one challenge by nearly all industry players—chemists, formulators, and managers alike. The resultant "reengineering" of performance presents myriads of new variables and just as many new questions. For instance, how can newly patented additive technologies and chemistries dramatically improve wetting and dispersing properties? How can the vast array of interfacial tension parameters be properly aligned and balanced? How can foam be alleviated? This text has been designed to answer these questions within both practical and theoretical frameworks. As such, this book is intended for use in the classroom, in industry, and in governmental/regulatory agencies. By no means whatsoever, though, is it implied that all issues are addressed within a scant 300 pages. Recent advances have so greatly expanded the realm of newly patented performance enhancement technologies that an encyclopedic twenty-volume series would obviously be required to exhaustively review every topic. As a result, only the most exciting reengineering techniques will be discussed in this introductory volume. This text is divided into five parts—Part I systematically identifies and describes the "missing link", or the common denominator, that can allow one to properly integrate all performance parameters; Parts II through IV provide detailed coverage of the most critical technologies—wetting and dispersing, interfacial tension (flow, levelling, and surface parameters), and defoaming. Part V concludes the text with integrative discussions of performance synergies and economics. To facilitate presentation, more than 380 special tables, graphs, flowcharts, formulations, and case studies are provided. Performance enhancement variables, along with concomitant coverage of both technical and economic concerns, are provided in an easy-to-use format equally suitable for use in the classroom, the laboratory, or the boardroom.

All materials in this textbook have been extensively "tried and tested", so to speak, in both classroom and seminar settings. Accordingly, the copious feedback and input received from more than 29,000 international attendees has helped define, channel, and guide the format of this book. Given the complexities of environmentally friendly coatings, selected topics will require explanatory prefaces in which theoretical background material is discussed; nevertheless, such material will be presented in an overview fashion so that adherence to a practical frame of reference can be maintained. Wherever possible—diagrams, descriptive figures, chemical structures, and practical examples will be presented. In selected cases, the breadth of the subject matter at hand will require the inclusion of special appendices in which important supplementary information is discussed at length.

The intent of this textbook is clarity and integration, so this means that abstruse or marginally useful information is avoided at all costs. Feedback from both industrial and academic reviewers has consistently indicated that core issues should be presented in an unadulterated, yet efficient fashion; as a further corollary, this means that an intentional departure from the possible "monotony" of paragraph after paragraph of text is implemented. For instance, not only are diagrams, flowcharts, and all the accoutrements of graphical presentation methodologies employed—but a host of additional teaching aids is also included. As an illustration of this point, the special question-and-answer approach utilized in Chapter 4 greatly simplifies the exceedingly complex and challenging subject matter of high molecular-weight wetting and dispersing agents (advanced products which simultaneously contain surface-active groups, basic/acidic structures, linking groups, polymeric building blocks, and pigment adhesion moieties). This textbook is formulated on the precept that technology and learning are active, evolving, and interactive processes, and as such, they should not become mired in the theoretical realm. Questions, answers, and real-life case studies are essential elements in the proper presentation of integrated technological concepts. In a nutshell, theory and practice become one.

The use of additive trade names has been strictly avoided; likewise, there has been absolutely no attempt to include lists of suppliers and their products, nor has there been any attempt to provide superfluous bibliographies and reference lists. (Selected "Guides to Further Reading" are included, however, where necessary and appropriate; experimental data, and concomitant background information—both theoretical and empirical—were generated in the author's laboratories. A combined total of more than 174,000 laboratory hours—equivalent to 87 man-years of laboratory time—is represented by the above.)

Copious student feedback has been highly instrumental in the design of this text. As mentioned at the onset, raw material lists, general textbooks, bibliographies, and related materials of an eclectic, non-integrated nature are quite abundant. It is now time for a change in perspective, so an integrated approach serves as the key focus of this textbook.

A concerted approach to creating and maintaining a healthy environment is precisely what industry, academia, and government expect from all players in the global economy. Environmentally friendly coatings play an important role in ensuring a healthy ecosystem, and additives are high performance products expressly designed to interact with and enhance the performance of virtually all ingredients in the coating system—whether such ingredients be pigments, resins, solvents, or otherwise. The role of additives is specifically that of bringing together all the individual components, in essence, providing *gestalt*, or a state of optimal performance and integration. But everyone knows that *gestalt*, in and of itself, is simply not enough; we live in a competitive world in which performance—whether integrated or not—must be translated into profits, so that is precisely why additives for environmentally friendly coatings are even more essential than ever. Additives contribute to "the bottom line". Furthermore, both performance and profits can be optimized after one has gained a proper understanding of the recent advances in additive chemistry. Armed with this knowledge, one can take the mystery out of environmentally friendly coatings, and meet the challenge of the twenty-first century and beyond.

Wallingford, Connecticut   *Edward W. Orr*

Acknowledgements

Copious thanks for input regarding the educational design and format of this textbook are extended to a number of individuals, educational institutions, and members of the industrial community. As mentioned in the preface, the format and choice of materials in this text are the direct result of feedback from presentations to over 29,000 individuals. For each and every presentation, constructive feedback was requested and received; accordingly, the author expresses thanks to the following individuals and organizations: Dr. Shelby F. Thames, Distinguished University Research Professor, Professor of Polymer Science, the University of Southern Mississippi; Mr. Michael Bell, Director Educational Services, Federation of Societies for Coatings Technology; Dr. Christine Strohm, Editorial Director, Carl Hanser Verlag; Dr. Edmund Immergut, Retired Editorial Director, Hanser Publishers; the late Mr. Abel Banov, former Editor of *Coatings World* and *Coatings World Report*; Mr. Dale Pritchett, Publisher of the two aforementioned periodicals; Dr. Robert Lochhead, Associate Professor and Chair, Department of Polymer Science, The University of Southern Mississippi; Dr. Jim Woo, Professor, Eastern Michigan University Coatings Research Institute; Dr. Taki Anagnostou, retired Professor, Eastern Michigan University Coatings Research Institute; Dr. Dane Jones, Professor, Chemistry and Biochemistry Department, California Polytechnic State University; the FSCT Constituent Societies in Chicago, Cleveland, Dallas, Detroit, the Golden Gate region, Houston, Kansas City, Los Angeles, Montreal, New England, New York, the Northwestern region, the Pacific Northwest region, Pittsburgh, the Rocky Mountain region, St. Louis, the Southern region, and Toronto; Mr. Francis Borel, Secretary-General, Federation d'Association des Technicians des Peintures, Vernis, Emaux, et d'Imprimerie de l'Europe Continentale (FATIPEC); and researchers and formulators at BASF Corporation (Germany, the United States, and Japan), The Sherwin-Williams Company (United States and Germany), Akzo Nobel Coatings, Inc. (Netherlands and the United States), DuPont Company, Inc., ICI Americas, Inc. (the United Kingdom and the United States), Degussa Corporation (Germany and the United States), The Valspar Corporation, PPG Industries, Inc., Dow Chemical Company (Japan), Morton Automotive Coatings, Ford Motor Company, the Association of Formulation Chemists, the American Chemical Society, ISO Central Secretariat, Nederlands Normalisatie-instituut, the Office of Air Quality Planning and Standards, and the United States Environmental Protection Agency.

Thanks are also extended to the following individuals of BYK-Chemie GmbH (Wesel, Germany) and/or BYK-Chemie USA (Wallingford, United States): Dr. Klaus Oehmichen, Wolfgang Zinnert, Jeffrey Converse, Robert McMullin, Karlheinz Haubennestel, Alfred Bubat, Leen van Dam, Dr. Wolfgang Kortmann, Manfred Knospe, Dr. Wilfred Scholz, Wilhelm Klammer, Herbert Becker, Bernd Dawid, János Hajas, Axel Woocker, Heiko Juckel, Wilhelm Wessels, Heinz Utzig, Frederick

Lewchik, Gary Mallalieu, Patrick Lainé, Terry Lester, Eleanora Baribault, Christine Kisiel, Dawn Fassman, Carl Alessandro, Charles Collinson, Oliver Dixon, Richard Edgar, Eugene Franklin, Jerry Kelley, John Lawrence, Way Lin, Stuart Lipskin, Robert Loper, John Palermo, Phillip Reynolds, Suzanne Farnsworth, Philip Saglimbeni, Otto Schmidt, Scott Shier, Alex Vignini, and Stanley Walzak.

As mentioned in the preface, the evolutionary nature of the coatings industry is incredibly dynamic and ever-changing; this textbook—as the first installment of a multi-volume series on additives, pigments, resins, solvents, and end-use technologies—is by no means complete. Additional volumes will be designed to revise, augment, and complement this text.

Synopsis: Performance Enhancement in Coatings

Of all the challenges encountered in the coatings industry, performance enhancement—particularly in environmentally coatings—is consistently ranked as the number one challenge by nearly all industry players—chemists, formulators, and managers alike. Why is this? What makes environmentally friendly coatings so enigmatic, unique, and fascinating? After all, such coatings have risen to clear dominance in the industry, but they are undoubtedly the most difficult to formulate and the least understood of all systems.

This textbook is absolutely the first to describe how additives have evolved to not only take the mystery out of environmentally friendly coatings, but also to meet the challenge of the twenty-first century and beyond. Modern-day coatings have special needs in regard to rheology, film performance, and surface characteristics. For instance, a particular high solids coating may require special wetting and dispersing additives in order to match the performance of comparable low solids systems. In another high solids coating, interfacially active substances and defoamers may be necessary to maximize properties. Other new technologies (water soluble, latex, powder, etc.) pose similar challenges.

An overview of new additive technologies is presented. Special emphasis is placed upon the three most critical areas—wetting and dispersing, interfacial tension (flow, levelling, and surface parameters), and defoaming. The chemical determinants of additive performance are outlined for a variety of new coatings. Not only will structure-performance correlations be discussed, but also more than 380 special tables, graphs, flowcharts, formulations, and case studies will be displayed. Finally, there will be a discussion of the decision points necessary to determine the proper choice of additive.

Contents

Part I

The Common Denominator of Performance Enhancement

1 Introduction: New Generation Additives for the New Century

"Reengineering" and "Performance Enhancement" are the watchwords for the new century, and additives are the keys to success—especially in environmentally friendly coatings.

1.1 Reengineering and Performance Enhancement

The twenty-first century and beyond—what exactly awaits us in the coatings industry? How can we cope with the sometimes diametrically opposed constraints of technology and finance? In particular, as technology continues evolving toward environmentally friendly systems, how can both performance and economic considerations be properly integrated? In direct response to these issues, a series of exciting developments on the additive front has arisen. Such developments meet the emerging challenges of the new century by radically reengineering performance in the following three areas:

- Wetting and dispersing
- Surface and interfacial tension
- Defoaming and air release

The new era of environmental responsibility first ushered in by the 1970s shows no signs of relenting. Nowadays, absolutely no one is surprised by the steady crescendo in the usage of waterborne, high-solids, radiation-cure, powder, and hybrid systems. But the coatings industry is nonetheless faced with a quandary; change and evolution do indeed benefit the environment, but they can also play havoc with the bottom line. Behind all the sugarcoated rhetoric of reengineering stands a host of reformulation challenges and financial balancing acts. The solution: "New Generation Additives".

1.2 New Additive Technologies

As shown in the "New Generation Additives" column of Table 1.1, today's conventional additives must evolve on multiple fronts to solve the performance problems of the twenty-first century.

Table 1.1   New Additive Technologies for Environmentally Friendly Coatings

Coatings segment	Performance problems of the twenty-first century	New generation additives
High solids	• Pigment encapsulation • Flow • Rheology • Substrate wetting	• Urea thixotropes • Polymeric wetting and dispersing agents • Combination flow enhancers
Waterborne	• Micro-foam • Dispersion • Rheology • Substrate wetting • Stability	• Modified defoamers • Primary dispersants and wetting agents • Synergistic flow additives • Special thixotropes
Powder	• Substrate flow and levelling • Color matching • Film thickness	• Hybrid flow additives • Modified wetting and dispersing additives

Flow, rheology, wetting, dispersing, and interfacial tension parameters must be properly aligned and balanced. Advanced coatings are often united by one common factor—the need for high molecular-weight wetting and dispersing agents which can dramatically surpass the performance constraints of conventional additives—constraints formerly bounded by the limited ability of certain low molecular-weight polymers to properly accommodate the high surface tension ingredients (such as water) that are commonly encountered in environmentally friendly coatings. High molecular-weight polymers often contain 12 to15 pigment adhesion groups, thus enabling improved color development, rheological features, and pigment utilization (up to 60–70% less pigment may sometimes be required for the achievement of equivalent color development). This is reengineering at its best; both technological and financial constraints are optimized.

1.3 The Number One Challenge: Performance Improvement in an Integrated Fashion

The formulators of environmentally friendly coating systems are faced with an increasing array of stringent technical, environmental, and cost constraints. Many of the greatest challenges on the performance front are involved with wetting, dispersing, interfacial tension (flow, levelling, and surface parameters), and defoaming. How can such parameters be optimally controlled? The answers to this crucial question can be outlined, in an integrated fashion, by describing the proper methodologies to control interfacial tension.

1.4 Identifying the "Missing Link"

"Interfacial tension" is not only an important phenomenon; it is truly the "missing link"—the common denominator which can serve as the basis to describe precisely what happens at virtually all of the interfaces encountered upon the manufacture, application, and usage of modern coating systems. For example, one of many important interfaces occurs at the pigment/resin molecular interaction point (Fig. 1.1). Other germane factors and interfaces must also be considered within the context of solid particle chemistry and particle geometry, as shown in Figure 1.1. Pigment and resin design are, of course, very important methods of initial control, but after tailored pigment and resin design have been taken into account, specialized additives should be considered. Such additives can function at a variety of interfaces and can be used to facilitate production and application.

As one pursues a comprehensive study of environmentally friendly coatings, one must incorporate myriads of physicochemical, steric, and dynamic particle movement variables into the program of study. How do pigment particles interact with one another when placed in contact with various coating constituents? What are the determinants of resin/pigment/water (or solvent) interfacial tension values? How can one *simultaneously* control (1) wetting at the microscopic level of pigment molecules, and (2) wetting at the macroscopic level of the substrate? The answers to questions such as the above will be provided in subsequent chapters and will form the basis of a comprehensive study of wetting, dispersing, and surface control; the remainder of this introductory chapter will serve merely to set the stage.

Physicochemical, steric, and dynamic particle movement variables must be controlled and coordinated within the coating system so that the final coating

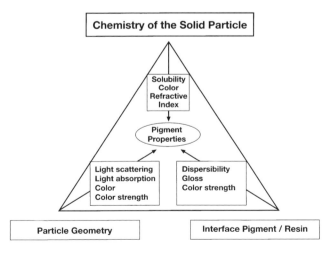

Figure 1.1 Pigment properties and interfaces

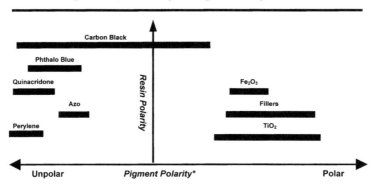

The Comparative Polarity Ranges of Pigment Surfaces

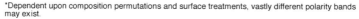

*Dependent upon composition permutations and surface treatments, vastly different polarity bands may exist.

Figure 1.2 Polarity

product is a completely *stabilized* system. Obviously, tailored stabilization systems must be provided according to coating type. As described in subsequent chapters, *significantly different* stabilization methods must be employed for water-based versus solvent-based systems. Within the context of stabilization methods, variables such as the pigment/surface polarity (Fig. 1.2) and the previously mentioned resin/pigment interface must be integrated into a map-like, multi-dimensional grid.

How many variables and dimensions would be required to truly describe the "inner life" of a coating system? How many formulator-influenced factors are there? Not every variable, dimension, and factor can be delineated within a scant 300 pages, but several dozen criteria will be introduced—more than enough to dramatically enhance overall coating performance; a partial list of key criteria is shown in Figure 1.3.

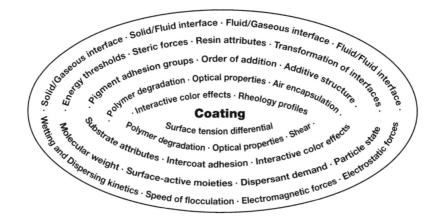

Figure 1.3 Key criteria

Indeed, Figure 1.3 could be expanded tenfold, but the objective of this text is not the presentation of superfluous lists; in contrast, the objective is the presentation of a series of integrated performance-improvement principles. This means that brevity is a virtue; accordingly, only the most important coating-related variables will be introduced. As much as possible, simple methods of controlling and classifying such variables will be employed. One such method involves the systematic examination of the surface tensions of substrates and coating system components (Table 1.2).

In order to optimize performance, the surface tension and interfacial tension values of all coating constituents (resins, pigments, and liquid carrier media) must be tailored to match the required demands of the substrate. As shown in Table 1.2, it comes as no surprise that the surface tension values of some of the more modern coating system components and substrates are higher than those of their more traditional counterparts. For instance, a difference of over 40 dynes/cm (please note that, in regard to semantics and nomenclature, 100 dynes/cm = 100 mN/m) exists when comparing water to many "traditional" solvents; and a difference of over 20 dynes/cm can exist between SMC-polyester and various phosphatized steel substrates. *Such surface*

Table 1.2

	mN/m
Solvents	
Water	72
2-Butoxy-ethanol (EB)	28
Butyl alcohol	23
Isopropyl alcohol	22
Propylene glycol methyl ether (PM)	28
Dipropylene glycol methyl ether (DPM)	31
Resins	
Acrylic latex	30–38
Alkyd emulsion	29–34
Alkyd solution	28–36
PUR emulsion	32–36
Melamine resin	42–58
PVAc latex	30–35
WR polyester	34–38
WR polyurethane	28–34
WR acrylic	32–38
Substrates	
Steel (phosphatized)	43–46
Steel (untreated)	25–30
Vinyl sheet	25–35
Propylene	28–30
PTFE	19
Polyester SMC	22–28

tension differences have formed a major impetus behind the design of many of the new molecular structures to be described in subsequent chapters.

1.5 The Interrelatedness of Wetting, Dispersing, Interfacial Tension, and Defoaming

Interfacial tension is the most important determining variable in the study of nearly all performance improvement variables. Whether the coating system is water- or solvent-based, UV-cured, or even powder-based; interfacial tension control is of utmost importance. Specialized coating additives are designed to function at "boundaries" and can therefore control interfacial tension.

Various categories of performance enhancement will be described—beginning with the dynamics of pigment wetting and dispersing—and followed by the challenges of surface phenomena and defoaming behavior. A step-by-step analysis of performance improvement methodologies will be presented throughout the text, with major emphasis being placed upon achieving reproducible results and significantly improved quality.

Judicious usage of additives improves performance through the coordination of physicochemical, steric, and dynamic particle movement variables. Nevertheless, the complexities of the above variables can once again be distilled down into one common theme—the control of interfacial tension within and throughout both the coating system and the coating system's contact points in the environment.

Through the combination of relatively new instrumental techniques, (e.g., Rodenstock Laser Surface Detector; an actual statistical analysis of Rodenstock-derived surface features is shown in Figure 1.4) and through the judicious usage of interfacially active additives, significant improvements can accrue. Such improvements occur not only in the realm of performance enhancement, but also in the realm of both cost reduction and improved customer satisfaction.

Laser Surface Detector

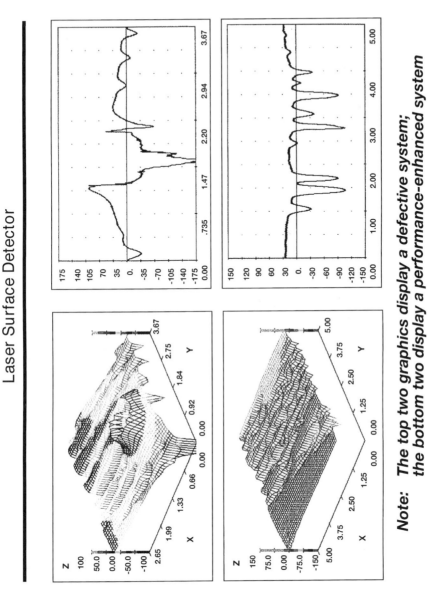

**Note: *The top two graphics display a defective system;
the bottom two display a performance-enhanced system***

Figure 1.4 Instrumental techniques

Guide to Further Reading

Alfke, G., Beyrau, M und Lehle, F., *Aktuelles Handbuch der Luftreinhaltung*, 1. Auflage (1983) Wilhem Jüngling GmbH & Co., KG, D-8047 Karlsfeld b. München. (3 Bände, Loseblattsammlung)

American Conference of Government Industrial Hygienists, *Documentation of the Threshold Limit Values for Substances in Workroom Air* (1971) Cincinnati, Ohio

American Conference of Governmental Industrial Hygienists, *Threshold Limit Values for Chemical Substances and Physical Agents and Biological Exposure Indices for 1990–1991*, Cincinnati, Ohio/ U.S.A. Bezug: ACGIH, 6500 Glenway Ave., Bldg. D-7 Cincinnati, Ohio, 45211-4438, U.S.A.

Bundesarbeitsamt, TRGS 900: Technische Regeln für gefährliche Arbeitsstoffe, *MAK-Werte* (1990) W. Kohlhammer Verlag GmbH

The Bureau of National Affairs, Inc., *Environment Reporter*, Washington, D.C., *International Environmental Guide*, Washington, D.C.

The Bureau of National Affairs, Inc., *Environmental Science and Technology*, American Chemical Society, Washington, D.C.

DFG Deutsche Forschungsgemeinschaft, "Maximale Arbeitsplatzkonzentrationen und Biologische Arbeitsstofftoleranzwerte 190", Bezug: VCH Verlagsgesellschaft mbH, Postfach 1260/1280, D-6940 Weinheim/Bergstraße

Dowd, E.J., *Air Pollution Control Engineering and Cost Study of the Paint and Varnish Industry*, Office of Air Quality Planning and Standards (1974) Research Triangle Park, North Carolina

Europäische Gemeinschaft, "Veröffentlichung des EINECS-Verzeichnisses (90/C 146/04)", Nr. C 146 S. 4 im Amtsblatt 15.06.1990

Gardon, J.L., Prane, J.W. (Eds.), *Nonpolluting Coatings and Coating Processes* (1973) Plenum Press, New York

Institut der Niedersächsischen Wirtschaft, Materialien 4, *Sonderabfallentsorgung in der Bundesrepublik Deutschland—Ein kurzer Leitfaden über Organisation und Struktur in den Bundesländern*, 3. überarbeitete Auflage, Stand: September (1989) Schiffgraben 36, D-3000 Hannover 1

Jacobs, M.B, *Analytical Toxicology of Industrial Organic Poisons* (1967) Interscience Publishers, Inc., New York

Noyes Data Corp., *Pollution Analyzing and Monitoring Instruments* (1972) Park Ridge, New Jersey.

Paint Research Association, *Health, Safety, Environmental Pollution and the Paint Industry* (1977) Middlesex, England

Patty, F.A. (Ed.) *Industrial Hygiene and Toxicology, Second Edition*, John Wiley & Sons, Inc., Vol. 1 (1958), Vol. 2 (1963) New York

Production Statistics/Produktionsstatistiken, *Farbe + Lack*, 9/1965, 9/1975, 5/1985, 7/1995, Editorial Reports

Quellmalz, E., *Das neue Chemikaliengesetz—Handbuch der gefährlichen Arbeitsstoffe* (1989) WEKA-Fachverlage GmbH, D-8901 Kissing. (4 Bände, Loseblattsammlung)

Roth, L., *Wassergefährdende Stoffe* (1988) ecomed Verlagsgesellschaft mbH, D-8910 Landsberg/Lech. (3 Bände, Loseblattsammlung)

Sax, I.N., *Dangerous Properties of Industrial Materials* (1975) Van Nostrand Reinhold Co., New York

Schildhauer, C., *Environmental Information Sources—A Selected Annotated Bibliography* (1972) Special Libraries Association, New York

Spence, J.W., Hanie, F.H., *Paint Technology and Air Pollution: A Survey and Economic Assessment* (1972) Office of Air Programs Publication No. AP-103, G.P.O., Washington, D.C.

Verchueren, K., *Handbook of Environmental Data on Organic Chemicals* (1977) Van Nostrand Reinhold Co., New York

Weinmann, W., Thomas, H.P, *Gefahrstoffverordnung mit Chemikalien-Gesetz* (1988) Carl Heymanns Verlag KG, D-5000 Köln. (4 Bände, Loseblattsammlung)

2 Interfaces and Evolution

2.1 Dissecting the "Missing Link"

Interfaces play an important role in coatings production and application. For the formulation chemist, the following interfaces are of special interest:

- Solid/gaseous
- Solid/fluid
- Fluid/gaseous
- Fluid/fluid

During the dispersion process, the solid/gaseous (pigment/air) interface is replaced by the solid/fluid (pigment/resin solution) interface. For substrate wetting, the solid/fluid interface plays an important role; in addition, foam is a highly increased fluid/gas interface; and one has a fluid/fluid interface when several paints come into contact as long as they are still wet and mobile (spray dust, wet-on-wet process, etc.). If one of the phases is gaseous, then the term "surface" is used; however, "interface" is the more general expression. In practice, though, this distinction is not always made.

2.2 Interfacial Tension

The forces which act at the interface are collectively called interfacial tension (surface tension). Such tension is often a property of the microscopic border-layer, where one can frequently observe change in density from one phase to the other. Without delving too deeply into theory, the forces displayed in Figure 2.1 will now be explained.

Interfacial Tension / Surface Tension

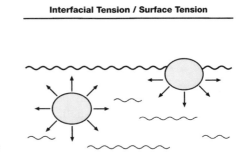

Figure 2.1 Forces at the interface

A typical molecule in a liquid is surrounded by other molecules which interact with the first one. In the bulk of the liquid, the resultant interactive forces are equal in all directions, so they balance each other. For a molecule closer to the surface, or directly in the surface layer, this is no longer valid though. Accordingly, those forces which are directed toward the bulk of the liquid (and which try to remove the particle from the surface) exert a dominant influence. Due to the influence of the surface tension, the liquid tends to maintain its surface as small as possible. Energy is required to increase the surface area. The interfacial tension is often named y and is reported as force (mN) per length (m); in many industry circles, however, the standard unit of measurement is still conventionally reported as dynes/cm. (Please note that 100 mN/m $=$ 100 dynes/cm.)

2.3 Increased Technical Demands Placed upon Additives

Given the evolution of coatings technologies toward more environmentally acceptable solvents, resins, and application/curing techniques; increased usage of interacially active substances is required. In bygone days, resin systems and solvent combinations often demonstrated quite low surface tensions and therefore made formulation easier in comparison to today's modern systems. For example, the more modern ingredients (and/or usage conditions) shown in column 2 of Figure 2.2 pose greater interfacial tension challenges than their predecessors in column 1. To cope with the challenges of advanced coating systems, additives have evolved in both structure and performance. An overview of new and integrated additive technologies is presented. Special emphasis is placed upon three areas—wetting and dispersing, surface flow and levelling, and defoaming (Fig. 2.3). The chemical determinants of additive performance are outlined for a variety of new coatings. Not only will structure-performance correlations be discussed in subsequent chapters, but also practical examples will be displayed. Finally, there will be a discussion of the decision points necessary to determine the proper choice of additive.

For years, managers in the coatings industry, not only in North America but also throughout the world, have been conditioned to believe that "change is the only constant", and that the future will be even more turbulent. So where does this leave the market for specialized additives? Thus far, one of the few trends consistently showing no sign of abatement has been the trend toward escalated environmental pressures. Concomitantly, new additive technologies often stand alone in being able to not only keep up with, but also surpass, the performance demands of environmentally friendly systems.

An encyclopedic twenty-volume treatise would not be long enough to describe all the complex economic and performance advantages imparted by the new additive technologies. In a nutshell, the shift toward additives providing integrated performance improvements appears, at least for the medium term, to be as robust a trend as the shift toward environmentally friendly products. Obviously, the

	1	**2**	**3**
● **"Solvents"**	Mineral Spirits	Water	Powder
● **Resins**	Alkyds	Polyesters	Epoxies
● **Application / Curing Tech.**	Brush / Air Dry	Spray / Bake	Bell / UV

Figure 2.2 Comparative evolution of coatings technologies

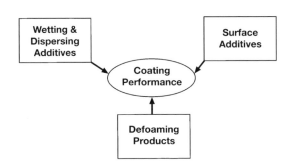

Figure 2.3 Performance evaluation in three realms

soothsayers of today may be wrong tomorrow, but at least where the new additive technologies are concerned, there exist myriads of reasons in support of the continued rise of specialized, high molecular-weight products. Even if the undeniably strong undercurrent of environmental pressures were not enough to fuel the drive toward advanced additives, for instance, integrated wetting and dispersing products with 12 to 15 pigment adhesion groups—there would still exist unequivocal evidence in favor of new and advanced technologies. Namely, the high molecular-weight additives improve cost and performance even in systems which are not environmentally friendly.

2.4 Stability for a Changing Market

International competition, reformulation pressures, and environmental trends may have squeezed profit margins for additive manufacturers, but new and innovative ways of enhancing performance are a welcome release—in effect, a safety valve providing a measure of stability in an ever-changing additive market. The most dependable way of mastering the future, at least in the medium term, is reliance upon the proven advantages of additives designed to reengineer performance.

(*Special Note:* For obvious reasons, certain graphic depictions [including, for instance, Fig. 2.1, Fig. 3.4, Fig. 3.7 and countless others] are not drawn to scale. In addition, it is important to state that a full description of variables, axes, and/or "dimensional idiosyncrasies" [including those involved with attractive and repulsive forces] would often require several dozen pages. Where appropriate, though, only the "practical endpoint" or "final result" is shown.)

Guide to Further Reading

American Chemical Society, *Advances in Chemistry Series 43* (1964) Washington, D.C.

Bikerman, J.J., *Surface Activity*, Second Edition (1958) Academic Press, Inc., New York

Bondy, C., Role of Surfactants in Emulsion Polymerization and Emulsion Paints, *J. Oil Colour Chem. Assoc.* (1966) 49 (12), p. 1045–1062

Diesel, E.W., Lühr, H.P., *Lagerung und Transport wassergefährdender Stoffe* (1982) Erich Schmidt Verlag, Berlin (4 Bände, Loseblattsammlung)

Elworthy, P.H., Florence, A.T., MacFarlone, C.B., *Solubilization by Surface-Active Agents* (1968) Chapman and Hall, Ltd., London

Kuehn-Birett/Ecomed, *Merkblätter Gefährliche Arbeitsstoffe*, 9. Auflage des Grundwerks (1990) Ecomed Verlagsgesellschaft mbH (7 Bände, Loseblattsammlung)

Moilliet, J.L.,Collie, B., Black, W.D., *Surface Activity*, Second Edition (1961) Van Nostrand Co., Inc., Princeton

Sisley, J.P., *Encyclopedia of Surface-Active Agents* (2 Vols.) (1952–1964) Chemical Publishing Co., Inc., New York

Somorjai, G.A., *Principles of Surface Chemistry* (1972) Prentice-Hall, Inc., New York

Springer,V., *Handbuch der gefährlichen Güter*, 4. Neubearb. Auflage (1987) Berlin (4 Bände, Loseblattsammlung)

Part II

Wetting and Dispersing
(The Control of Pigment Interfaces)

3 The Mathematical and Practical Determinants of Pigment Wetting and Dispersing Improvement

Overview—Because of the complexity of wetting and dispersing phenomena, this chapter introduces the prerequisites which are essential to developing an integrated knowledge base. Recent technological advances have allowed researchers to more adequately define the mathematical determinants of wetting and dispersing. Fortunately, *such advances have progressed far beyond the rather esoteric theoretical realm of "mathematical equations"; **tailored** performance improvement of a very practical nature can now be achieved* through the utilization of either specially modified formulating procedures or newly patented additive chemistries.

A wide variety of performance enhancement methodologies will be discussed within the context of the following focal points:

- Introductory wetting and dispersing kinetics
- Speed of flocculation
- Special energetic effects (electromagnetic and electrostatic forces)
- Overcoming energy thresholds: Winning the "dispersion battle"
- The control of steric forces
- Practical concerns
- A synopsis of ten integrated performance improvement principles

Conceptually speaking, the chemical and physical determinants of wetting and dispersing can be overviewed within a simplified mathematical framework. The utility of the resultant mathematical paradigm lies primarily in its *reproducible applicability to a wide variety of practical formulations*; nevertheless, at least one "definitional caveat" should be mentioned before any discussion of wetting and dispersing phenomena can begin. Beyond a shadow of doubt, a multivolume treatise would not suffice to cover every aspect of the above subject matter; accordingly, the prime intent of the overview publication at hand is the introduction of the major focal points mentioned above—in particular, the overview of half a dozen specially selected topics which prove to be of particular interest and utility to the formulator of environmentally friendly coatings, inks, plastics, adhesives, pharmaceuticals, and other particle-containing liquid media.

Within the framework of the above discussion, the first portions of this chapter present classical introductory kinetics concepts along with more advanced "speed-of-flocculation" parameters. Special energetic effects (electromagnetic and electrostatic phenomena), in combination with steric forces, form the focal points of the middle of the chapter. Finally, to complete the picture, practical concerns, along with a synopsis of ten integrated performance improvement principles, will then be discussed at the end of the chapter.

3.1 Introductory Wetting and Dispersing Kinetics

The traditional "wetting and dispersing" process can actually be divided into three somewhat discrete and, from a mathematical perspective, easily identifiable phases as shown in Table 3.1. In order to set the stage for the ensuing discussions, a preview of selected aspects of the aforementioned three phases is now essential.

Table 3.1

Phase	Description
Wetting	Removal of adsorbed air (including, in some cases, adsorbed water vapor) from the pigment surface; subsequent formation of a pigment/resin interface
Dispersing	Mechanical "dispersing" or "breaking" of any and all pigment agglomerates which may be present (Provision of either a 100% primary particle state or a "controlled flocculation" state)
Stabilizing	Hindrance of reversion to a reagglomerated or uncontrolled flocculation state

3.1.1 Wetting

Thermodynamically speaking, wetting can take place only when the transformation from a pigment/air interface to a pigment/resin interface results in the attainment of a lower energy state. Unfortunately, one must generally overcome several energy thresholds before one can achieve the required lower energy level. From a kinetics perspective, the wetting process itself depends (according to one permutation of the classical Washburn equation shown in Figure 3.1) upon a variety of factors—including the viscosity of the resin solution, as well as the lengths and radii of the pigment pores. Although the Washburn equation provides a useful springboard for an introduction to mathematical determinants, it serves only as an illustrative starting

Washburn Equation

$$V = K \cdot \gamma_{Fl} \cdot \cos\theta \cdot \frac{r^3}{l \cdot \eta}$$

v = Wetting speed
K = Constant
γ_{Fl} = Surface tension of the liquid phase
θ = Contact angle
r = Radius of the pores
l = Length of the pores
η = Viscosity of the liquid phase

Figure 3.1 Introductory wetting kinetics equation

point and will, of course, have to be augmented by additional equations and commentary. The mathematical expression, $(\gamma_{FI} \cdot \cos \theta)$, represents "spreading pressure" or, for all practical purposes, the prime formulator-influenced factor in the expanded equation above. Interfacial tension modification (through enhanced formulating procedures or newly patented additive chemistries) provides the key to controlling the crucial $(\gamma_{FI} \cdot \cos \theta)$ factor.

In terms of its diagnostic utility to the coatings formulator, the "spreading pressure" can serve as an important tool in ascertaining why, for instance, a particular formulation may display suboptimal wetting properties. A common reason for improper wetting is the existence of a very large interfacial tension gradient (perhaps between a pigment's water/vapor envelope and a rather unpolar resin).

3.1.2 Dispersing and Stabilizing

The dispersing or grinding process generally involves the introduction of mechanically-derived shear forces. For instance, the rotational energy (imparted by the dispersion equipment) should theoretically permeate all portions of the liquid coating media. In practice, however, a host of intervening factors can foster flocculation and therefore hinder proper dispersion. Given the dependence of the aforementioned factors upon the speed of flocculation kinetics, the next section of this chapter will begin by elaborating upon some of the more important variables which influence speed and efficiency of dispersion and/or flocculation.

The stabilization process involves the prevention of "energy-state reversion," so to speak. Of prime importance is the "building of force fields of varying natures" (electrostatic/steric/physical/chemical) around either the idealized primary particle or the controlled flocculation entity. Additional discussion of the more important force field methodologies will serve as the focus of the concluding portions of this chapter.

3.2 Speed of Flocculation

Given the theoretical assumption that the thermally induced currents and/or movements (of primary particles) will lead to a localized "microflocculation" state *whenever two primary particles collide*—then the two special equations in Figure 3.2 are applicable*. In and of itself, Figure 3.2 provides a fascinatingly simple and useful overview of the rather straightforward theoretical relationships among time, viscosity, temperature, and "particle distribution density". One can quite easily calculate

---

*Because of the intentional brevity of this particular publication, the multipage mathematical background behind all theoretical assumptions will, of course, not be included.

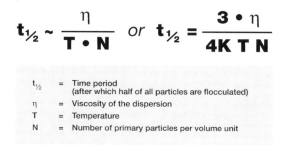

$$t_{1/2} \sim \frac{\eta}{T \cdot N} \quad or \quad t_{1/2} = \frac{3 \cdot \eta}{4K\,T\,N}$$

$t_{1/2}$	=	Time period (after which half of all particles are flocculated)
η	=	Viscosity of the dispersion
T	=	Temperature
N	=	Number of primary particles per volume unit

Figure 3.2 Speed of flocculation calculations

primary particle behavior and the resultant flocculation kinetics; nevertheless, there exists one very important problem for the practitioner (namely, the formulator, in this case). The aforementioned equations are excellent for utilization as predictive indices of *microflocculation* behavior; the real world, though, involves *macroflocculation* phenomena. If one were to depend on merely the above equations for practical results, then one would constantly be faced with the same dead-end dilemma. According to both of the above equations, all macroscopic dispersion states inevitably result in flocculation. *Obviously, the above theoretical conclusion is incorrect, so a series of important additional variables must also be factored into the above "theoretical equation scenarios".* Such variables form the focal points of the following discussions.

3.3 Special Energetic Effects (Electromagnetic and Electrostatic Forces)

Generally speaking, electromagnetic (attractive) and electrostatic (usually repulsive) forces are *interpigment phenomena* which must be controlled as crucial elements of the "first line of flocculation defense". (Additional defense elements include steric forces and special combination effects; however, these topics will be discussed later.)

Because of the often pairwise occurrence of electromagnetic and electrostatic forces, one cannot adequately discuss one force type without at least introducing the other. Logic dictates that interparticle attractive forces must be overcome first; accordingly, electromagnetic phenomena will be overviewed as a prelude to the ensuing discussion of electrostatic effects.

3.3.1 Electromagnetic Forces

First of all, what exactly are the major electromagnetic (and/or nonelectromagnetic) forces which commonly occur in coatings systems? Naturally, a preliminary listing of

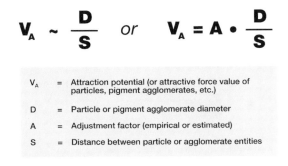

$$V_A \sim \frac{D}{S} \quad or \quad V_A = A \cdot \frac{D}{S}$$

V_A	=	Attraction potential (or attractive force value of particles, pigment agglomerates, etc.)
D	=	Particle or pigment agglomerate diameter
A	=	Adjustment factor (empirical or estimated)
S	=	Distance between particle or agglomerate entities

Figure 3.3 Electromagnetic force variables

such forces might begin by including countless phenomena—such as modified hydrogen-bonding, dipole, and solubility variables; however, *for the purposes of a practice-oriented overview, the electromagnetic force variables displayed in Figure 3.3 provide a more beneficial avenue of approach.*

From a mathematical perspective, one can employ the aforementioned variables to discern the relatively simple trends graphically depicted in Figure 3.4. In practice, the obvious implications (and concomitant advantages) of small versus large particle size pigments can often be controlled at will by the formulator.

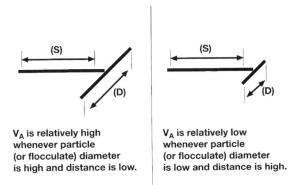

V_A is relatively high whenever particle (or flocculate) diameter is high and distance is low.

V_A is relatively low whenever particle (or flocculate) diameter is low and distance is high.

Figure 3.4 Graphic depiction of electromagnetic force trends

3.3.2 Electrostatic Forces

The composition of a typical coating formulation is conducive to the occurrence of both attractive forces (as discussed previously) and repulsive forces. Electrostatic phenomena can occur as a result of electrical charges on the pigment particle surfaces.

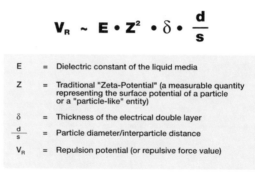

Figure 3.5 The magnitude of pigment repulsive forces

To be more exact, the repulsive forces are related to the thickness of the classical electrical double layer (δ). Because the structure of the double layer itself is composed of a "cloud" of counterions aggregated around an already charged pigment particle*, then the induced proximity of two such "clouds" results in the formation of repulsive forces *between* the respective clouds. The magnitude (V_R) of the aforementioned repulsive force can be calculated as shown in Figure 3.5.

Furthermore, the *thickness* of the electrical double layer (Fig. 3.6) depends upon the dielectric constant of the liquid media *and upon both the concentration and the "ionic value" of the electrolyte at hand.*

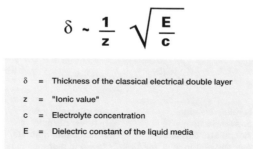

Figure 3.6 Thickness of the electrical double layer

It is interesting to note that the intentional or unintentional addition of electrolytes can indeed cause flocculation; the δ value is lowered, and the concomitant repulsive potential becomes weaker. The crucial practical implications of the above become even more apparent *when one considers the additional fact that the degree of flocculation correlates with the ionic value raised to the **sixth** power!* In other words, certain types of

*Or perhaps even around a pigment flocculate.

"impurities" or pigment treatments (such as those containing iron or calcium moieties) can play havoc with the formulation. Undoubtedly, this is no surprise to the experienced formulator; nevertheless, the fact that various "impurities" can exert such vastly different influence levels is certainly worth closer examination. Consider the comparative situation that exists with sodium (Na^+) versus calcium (Ca^{2+}) versus iron (Fe^{3+}) ions. The differential influence levels which could be exerted by the above ions range from one (1^6) to sixty-four (2^6) to seven hundred twenty-nine (3^6)! Of course, one can utilize the above information in his choice of pigments, pigment treatments, and formulation tactics. Quite surprisingly, the ubiquitous influence of certain contaminants is often overlooked; an excellent case in point is the choice of water, solvent, and/or storage and delivery vessels. Numerous instances of documented dramatic dispersion improvement have occurred as a direct result of the mere elimination of routine contaminants in either the liquid media or in the associated vessels.

3.4 Overcoming Energy Thresholds: Winning the "Dispersion Battle"

Thus far, the independent influences of both attractive forces (V_A) and repulsive forces (V_R) have been delineated; however, the prime point of importance to the coatings formulator is generally the unequivocal dominance of the repulsive (dispersion-stabilizing) forces. In other words, the "force field summation" of V_A and V_R must result in *an overall preponderance of repulsive forces, as demonstrated in the shaded zone of Figure 3.7*. The shaded zone in Figure 3.7 is of special importance to the achievement of dispersion stability; *proper dispersion will occur **only** when the repulsive forces predominate and when the interparticle spacing is at least as great as the critical distance (S_{CR}).*

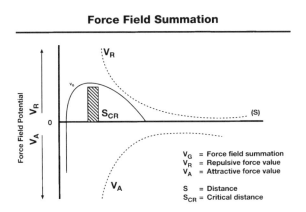

Figure 3.7 Overcoming energy thresholds through the summation of forces

3.5 The Control of Steric Forces

Definitionwise, ideal steric hindrance provides stable interparticle spacing through the utilization of specially designed additives which adsorb onto the particles; furthermore, such additives unfurl a series of voluminous, protruding molecular chains into the resin solution. The end result is ideal pigment spacing along with the virtual elimination of destabilization effects. Optimal color development and completely flood/float-free systems can be easily achieved.

For decades, traditional solvent-based coatings have generally employed low to moderate steric hindrance as a dispersion stabilizing technique in solvent-based systems. Newly patented additive chemistries, however, have *not only increased stabilization strength by up to 900%, but have even allowed steric hindrance utilization in water-based systems.* The accrued advantages include not only performance improvement, but also substantial cost savings.

Obviously, an in-depth discussion of advanced steric hindrance techniques is far beyond the scope of this overview publication; however, two very important concepts must be emphasized. *First of all*, the key to improved performance generally involves the tailored usage of additive molecules which contain more than the usual one or two "pigment adhesion groups" (molecular entities which allow proper additive adsorption onto the pigment surface). As displayed in the top half of Figure 3.8, proper interpigment spacing may require not only optimal unfurling, but may also require up to a *dozen special pigment adhesion groups for some pigments. Second*, as displayed in the bottom half of Figure 3.8, improper spacing will inevitably result when the voluminous molecular chains of the additive *do not properly unfurl*. In practice, this means that one must be especially careful to choose the correct additive. Given the wide palette of newly patented additive structures (with tailored numbers and types of pigment adhesion groups), specially designed additives are now available for nearly all solvent-based and water-based formulations.

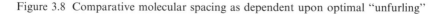

Provision of proper spacing through multiple "pigment adhesion groups" (12 in this case) and through optimal "unfurling"

Improper spacing through suboptimal "unfurling" and through insufficient resin solution interaction

Figure 3.8 Comparative molecular spacing as dependent upon optimal "unfurling"

3.6 Practical Concerns

Although the mathematical determinants of pigment wetting and dispersing are indeed important, the real issue of prime interest to the coating formulator is the translation of theory into practice, namely, proper control of viscosity, flow and handling properties, and the degree of flocculation or deflocculation. The structure of wetting agents which contain both (1) specialized pigment adhesion groups, and (2) pigment/resin/solvent-specific polymer chains can play a decisive role in controlling rheological behavior. Steric hindrance and electrostatic charge repulsion, when properly employed, can serve to tailor coating system response. Specialized wetting and dispersing additives such as those shown in Figure 3.9 can help control not only rheological performance, but also gloss, hiding power, durability, and the development of precisely specified hue characteristics.

Deflocculating additive structures include one or more pigment affinic groups and a specified number (depending on the exact performance desired) of resin-like chain structures. In order to function properly, the pigment affinic groups must be arranged closely together and should be situated in a relatively restricted total volume of molecular space. When such additives adsorb upon the pigment surface, the deflocculation state itself is then stabilized.

Controlled flocculation additives demonstrate several differences in relation to the above. The pigment affinic groups are not restricted to such a small volume of molecular space. In contrast, such groups can be widely distributed over the molecule and can function as bridges between different pigment particles. With the help of the additive, three-dimensional "controlled flocculate" structures are formed. The size and stability of such specialized flocculates are determined by the properties of the additive. Two important determining factors are (1) the interaction between additive structures, and (2) the interaction between additive and pigment structures. On a microscopic basis, the resultant state of controlled flocculation is shown in the middle of Figure 3.10.

Structure of Wetting and Dispersing Additives

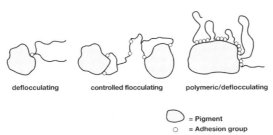

deflocculating   controlled flocculating   polymeric/deflocculating

◯ = Pigment
○ = Adhesion group
⌒ = Polymer chain

Figure 3.9 Comparative structures

Flocculation / Deflocculation

Flocculation   Deflocculation

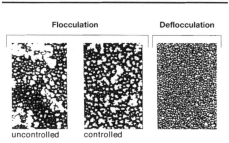

uncontrolled   controlled

Figure 3.10 Micrographs of flocculation/
deflocculation states

Controlled Flocculation

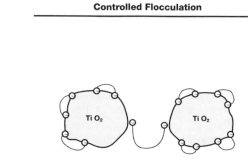

Figure 3.11 The intermediary nature of the
additive

It is essential to distinguish between "controlled" and "uncontrolled" flocculation. Without the presence of the additive, uncontrolled flocculation can develop and allow pigment particles to remain in direct contact with one another. However, in the presence of the additive, no direct pigment/pigment contact occurs. As shown in Figure 3.11, the intermediary nature of the additive functions as a spacing mechanism between individual pigment particles.

The intermediary nature of the above and similar additive structures can be achieved through several different methods. Steric hindrance, electrostatic charge repulsion, and "pigment adhesion group" to "pigment adhesion group" contact can all play important roles in controlling degree of flocculation.

3.7 Deflocculating Additives

Through utilization of deflocculating structures, a rather Newtonian flow behavior is generally created. In addition, the viscosity is often reduced. This provides the further advantages of improving flow properties and color development. Features such as gloss, transparency, and/or hiding power* may also benefit from the small particle size that is achieved.

Deflocculating additives can assist in equilibrating the mobility of pigment particles of disparate size and shape. For instance, inorganic and organic pigment structures (as shown in Fig. 3.12) can be acted upon by the additive in such a way that *all pigment species display equivalent mobility features.*

The equivalent mobility mentioned above can be created by forming, for instance, a special "interactive molecular sphere" around the organic particles. Such a "sphere" is shown in the right-hand portion of Figure 3.13. The point of importance is that the

*Of course, whether or not transparency versus hiding power is improved can be, in part, dependent upon the original intent (transparency or opacity) as specified by the pigment manufacturer.

Pigment Structures   **Movement of the Pigment**

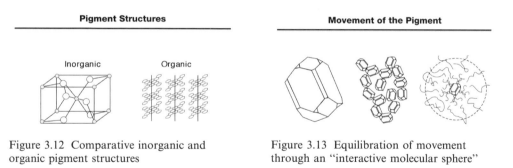

Figure 3.12 Comparative inorganic and   Figure 3.13 Equilibration of movement
organic pigment structures   through an "interactive molecular sphere"

small organic pigment particle in the middle of the interactive sphere is affected by the additive in such a way that the organic pigment particle mobility matches the mobility of the inorganic pigment particle. Often, the term used to refer to such complete flocculation is "ideal dispersion". (The left-hand portion of Fig. 3.13 shows the inorganic pigment particle for comparative purposes.) Obviously both steric hindrance and electrostatic charge repulsion can aid in adjusting mobility. Molecular weight of the additive itself is also an important consideration. Other factors which should be incorporated into the design stage of the additive include polarity, compatibility with polar and non-polar solvents, suitability for separate-grind usage, suitability for co-grind usage, etc. For instance, what happens when the molecular weight of the additive is increased? Suitability for use in separate grinds increases along with polarity. Post-addition for the higher molecular weight species is not recommended. On the other hand, the lower molecular-weight products are more suitable for co-grinds and can, in certain special cases, even be post-added. A diagram summarizing various design features of one family of additives is shown in Figure 3.14.

Polymeric Wetting and
Dispersing Additive Differences

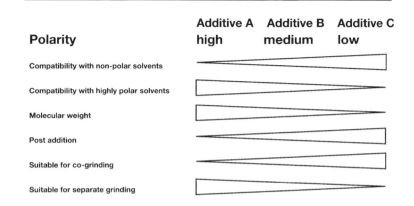

Polarity	Additive A high	Additive B medium	Additive C low
Compatibility with non-polar solvents			
Compatibility with highly polar solvents			
Molecular weight			
Post addition			
Suitable for co-grinding			
Suitable for separate grinding			

Figure 3.14 Design features

3.8 Controlled Flocculating Additives

The judicious usage of controlled flocculating additives leads to the formation of special three-dimensional structures. Thixotropic flow behavior can be brought about by such structures. In a resting state, the viscosity of the paint may be rather high; however, when subjected to shear forces, the structures (controlled pigment flocculates) break apart and therefore provide lower viscosity. Afterwards, when the shear forces cease, the controlled structures slowly rebuild.

Deleterious sagging behavior can often be alleviated through the usage of controlled flocculating structures. During processing, the paint viscosity is low enough to assure ease of handling. Afterwards, in the resting state (e.g., following application), the viscosity increases so that macroscopic layer stability results, even with relatively thick layers upon vertical substrates.

The same positive stabilizing effect can also be observed in settling behavior. As a result of the higher viscosity encountered during the absence of shear forces (in this case referring to storage), the "settling speed" of the pigment particles is radically reduced through *controlled* flocculation. (Another important "settling factor" is also worthy of mentioning at this point; definitionwise, *un*controlled flocculation is the formation of very hard, compact sediments which are generally impossible to fully reincorporate. In contrast, *controlled* flocculation interjects additive molecules between the pigments and will not permit the formation of hard, compact sediments.)

The prevention of settling requires not only the absence of direct pigment/pigment contact, but also requires the presence of specialized pigment adhesion groups within the molecular structure of the wetting and dispersing additive itself. In the case of controlled flocculating additives, one unique feature of the pigment adhesion groups is that they have the capability of interacting not only with the pigments, but also with other pigment adhesion groups within the system. This is the key to proper control of settling and sagging behavior.

Comparative settling phenomena are shown in Figure 3.15; for illustrative purposes, Figure 3.16 displays the generalized effect of gravity in reference to vertical surface sagging.

Settling: Localized Rheological Differences
Flocculated Pigment Agglomerates

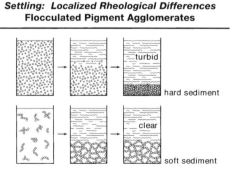

Figure 3.15 Comparative settling phenomena

Gravity Effects: Localized Rheological Differences
Macroscopic Levelling, Flow and Sag

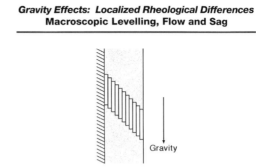

Figure 3.16 The effect of gravity in reference
to vertical sagging

3.9 A Synopsis of Ten Integrated Performance Improvement Principles

The mathematical and practical determinants of pigment wetting and dispersing improvement are indeed numerous. In addition, they may sometimes seem rather complex upon first glance; nevertheless, ten relatively simple and practical principles (as shown below) can be gleaned from the overview information presented thus far:

1. The traditional pigment wetting and dispersing process is actually divided into three somewhat discrete phases—wetting, dispersing, and stabilizing. Each phase can be described within the context of kinetics and/or energy thresholds; furthermore, performance in each phase can be enhanced through *the tailored control of interfacial tension*.

2. "Wetting" is the first phase and involves the transformation of a pigment/air interface into a pigment/resin interface. Both the *viscosity* and the *contact angle* are critical formulator-influenced factors in the more complex equation describing wetting behavior. Interfacial tension modification provides the most efficient formulating approach.

3. "Dispersing" is the second phase and involves the introduction of mechanically-derived shear forces to break any and all pigment agglomerates; either primary particles or controlled flocculates are the result. Formulator-influenced factors (from a microflocculation perspective) include *time, viscosity, temperature, and "particle distribution density"*.

4. *Micro*flocculation and *macro*flocculation are different; the latter more closely describes "real world" wetting and dispersing situations encountered by the formulator. A proper description of macroflocculation requires the inclusion of at least two additional formulator-influenced variables—*attractive* forces (V_A) and *repulsive* forces (V_R). Control of macroflocculation states is essential during both the dispersing and stabilizing phases.

5. "Stabilization" is the third phase and involves the prevention of "energy state reversion". Force fields of varying natures (*electrostatic/steric/physical/chemical*) are crucial.

6. Many attractive forces (V_A) can be subsumed under the main category of *electromagnetic forces*. The most pertinent formulator-influenced factors include *particle diameter* and the *distance* between the particles.

7. Two major types of repulsive forces (V_R) are *electrostatic* and *steric hindrance* phenomena. Both can be dramatically influenced not only by interfacial tension control (easily achieved with additives), but also by such factors as raw material selection and/or dispersion technique.

8. In order to win the "dispersion battle", the force-field summation of attractive forces (V_A) and repulsive forces (V_R) must result in an overall preponderance of repulsive forces.

9. The formulator's objective is always *to achieve a level of interparticle spacing which matches or exceeds the "critical distance"* (S_{CR}). Once again, this objective can be enhanced by proper interfacial tension modification.

10. Some of the more *advanced* performance enhancement techniques (such as steric hindrance produced by *multiple pigment adhesion groups*) can produce truly dramatic stabilization improvement along with substantial cost savings.

Given the ever-increasing importance of performance improvement, one can rest assured that wetting and dispersing phenomena will continue to serve as the focal point of ongoing research and development projects. Accordingly, the mathematical and practical determinants of pigment behavior will become even more closely integrated over time.

(*Special Note*: The author has observed the occurrence of an absolute plethora of different spellings, punctuations, and unit-to-unit conversions for certain technical terms and chemical nomenclature expressions in the industry. Often even the accepted industry reference books and dictionaries differ on the issues of "spelling variants", abbreviation shortcuts, hyphenation/punctuation rules and word-division protocols. Accordingly, several alternate technical terms, word spellings and grammatical usages—all equally correct—may indeed be employed by some readers.)

Guide to Further Reading

Bell, S.H., *J. Oil Colour Chem. Assoc.* (1952) 35, p. 386

Bode, R., *Kautschuk und Gummi*, Kunststoffe (1969) 22, p. 167

Boehm, H.P., *Farbe und Lack* (1973) 79, p. 419

Degussa-Pigmente für Druckfarben, Pigmente Nr. 10, überarb. 5. Aufl. (1977); Degussa, Frankfurt

Degussa-Produkte für Lacke und Farben, Pigmente Nr. 17, 2. Aufl. (1970); Degussa, Frankfurt/M.

DIN 5033 Blatt 1 (1970), DIN 53204 (1968), ASTM D 2414-70

Dulog, L., Schmitz, O., *Polymeradsorption und ihr Beitrag zur Stabilsierung von Pigmentdispersionen* (1994) Forschungsinstitut für Pigmente und Lacke e.V., Stuttgart, Institutsbericht, Nr. 14

Haselmeyer, F., Oehmichen, Dr. K., *Der Einfluß der Struktur des Netzmittels auf die Wirkungsweise und filmtechnischen Eigenschaften am Beispiel eines Alkydharzes* (1968) Fatipec

Honak, E., VI. *Fatipec-Kongreßbuch* (1962) p. 76

Jebsen-Marwedel, H., *Farbe und Lack* (1960) p. 314

Jettmar, W., *100 years BASF* (1965) p. 207

Koberstein, E., Lakatos, E., Voll, M., Ber. Buns, *Ges. Phys. Chem.* (1971) 75, p. 1106

Kresse, P., *Farbe und Lack* (1966), p. 111

Orr, E.W., Proceedings of the Association of Formulation Chemists, Las Vegas, September 24 (1997)

Patat, F., Killmanns, E., Schliebener, C., *Fortschritt der hochpolymeren Forschung* (1964) 3, p. 332

Parfitt, G.D., *Dispersion of Powders in Liquids* (1981) Applied Science Publishers, London

Schumacher, W., *Farbe und Lack* (1974) 80, p. 290

Schumacher, W., XIII. *Fatipec-Kongreßbuch*, (1976) p. 582

Studebaker, M.C., Huffman, E.W.D., Wolfe, A.C., Nabors, L.G., *Ind. Eng. Chem.* (1956) 48, p. 162

4 Converting Theory into Practice: The Transformation of Interfaces

4.1 An Overview of Advanced Wetting and Dispersing Technology

Armed with an understanding of the previously presented "Mathematical and Practical Determinants of Pigment Wetting and Dispersing Improvement," one is now ready to convert theory into practice. Consequently, the next logical step involves a brief review of selected principles, followed by a question-and-answer introduction to advanced wetting and dispersing technology. In addition, two special appendices (with more than 200 laboratory-tested formulations and 250 pigment-additive combinations) are included in support of this chapter.

As shown in Figure 4.1, solid particle media (such as pigments and fillers) must be capable of enduring the transformation from a solid/gaseous interface to a solid/liquid interface. Homogeneous particle distribution enhances the transformation process and provides the key to improved performance; interfacially active substances, which bridge the boundaries between solid and liquid media, are therefore essential. From a practical standpoint, proper distribution is required not only during production and application, but also upon storage.

4.2 Four Types of Performance Enhancers

As one would logically expect, both deflocculating and controlled flocculating additives may be necessary. In addition, rheology modifiers and/or thixotropic agents can play integral roles in the production of homogeneous pigment distribution states.

The Transformation of Interfaces

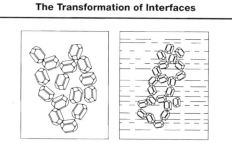

Figure 4.1 Solid particle media in transformation

Table 4.1

Lower molecular-weight polymers
- Based primarily on fatty acid chemistry
- Reaction products of unsaturated fatty acids with polyamines, polyalcohols, etc.
- Formation of polyamides, polyesters, and salts thereof

High molecular-weight polymers
- Special reaction products of polyisocyanates and/or related products with polyesters and/or polyethers (in combination with basic or acidic groups which provide powerful and stable adsorption layers on pigments and fillers of different structures)

Rheology modifiers
- Thixotropy increase by formation of three-dimensional frameworks with SiOH groups and/or OH-containing molecules
- Structures can be modified for tailored organic polymer compatibility

Liquid thixotropic agents
- Organo-modified ureas in polar aprotic solvents
- Formation of strong H-bonding forces when diluted in solvent systems
- Organic modification with high variability from low to high polarity

An introductory overview of selected developments in the above four additive categories is presented in Table 4.1.

Within the context of performance reengineering, each of the above additive categories is important; nevertheless, the lion's share of the most exciting developments has occurred on the high molecular-weight polymeric additive front. In fact, dramatically enhanced pigment utilization, color development, and/or reproducibility often result from the employment of newly patented polymeric additives. In some cases, up to 60–70% less pigment may be required for the achievement of equivalent color development. Accordingly, specialized high molecular-weight products, as displayed in Figure 4.2*, will serve as the primary focal point.

4.3 Questions and Answers about Performance Improvement

Given the practical nature of the following question-and-answer approach, an intentional departure from traditional "paragraph-style" text will be implemented. Ten crucial questions (as displayed in Table 4.2) will be posed and then systematically answered.

---

*Molecular structure is, of course, only one of many important variables; several dozen additional "wetting and dispersing variables" serve as the focal point of selected chapters in this textbook.

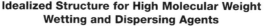

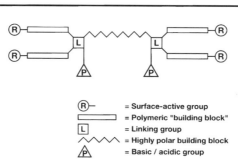

Figure 4.2 Idealized structure

Table 4.2

A. Why were high molecular-weight (HMW) wetting and dispersing additives developed?

B. What advantages do high molecular-weight products offer?

C. Exactly when is it possible to improve the reflocculation behavior of pigment "X" with a member of the high molecular-weight polymeric family? In other words, is there an easy-to-use, preliminary decision criterion which can help determine which additive is most appropriate?

D. After one has employed a special pigment/additive cross-reference chart, what is the next step in the "decision analysis"?

E. Why are relatively high dosages of HMWs (especially when used in conjunction with organic pigments) sometimes necessary? Furthermore, from an economic perspective, does the magnitude of the resultant performance enhancement really warrant usage?

F. How can one empirically examine (*with an easy-to-use methodology*) primary-particle, controlled-flocculation, and/or uncontrolled flocculation states?

G. What might happen if one were to add too much HMW additive?

H. Can pigment concentrates and/or full coating systems manufactured with HMW products be mixed with systems which do not contain HMW moieties?

I. Can HMW products be "post-added"?

J. Suppose that one has no flocculation problems with his traditional, "tried-and-true" grinding systems; should one still consider HMW products?

Question A: Why were high molecular-weight wetting and dispersing additives developed?
A series of four critically important market and technology trends (as shown below) served as the prime impetus behind the development of high molecular-weight (HMW) advanced products.

- *Environmental Legislation*
1. The transition from conventional to high solids and water-borne systems.
2. "Classification directives" and the prohibition of certain coating system ingredients.

- *Increased "Reengineering" and Economic Rationalization Efforts*
1. The evolution away from selected co-grinds to individual pigment grinds.
2. Pigment concentrates and mixed paint systems—40 base colors—10,000 mixed colors (in automotive refinish, for instance).

- *The Increased Challenges of Poorly Wetting Resins and Difficult-to-Disperse Pigments*
1. High energy costs for dispersion to reach optimal color strength.
2. Extensive testing required.
3. Change of color strength during shelf-life often possible.

- *The Growing Utilization of Pigment Concentrates*
1. Problem solving by "third parties"
2. Ready-made concentrates.

Question B: What advantages do high molecular-weight products offer?
The primary utility of the aforementioned additives is their ability to efficiently and effectively function with nearly all pigment types. Of course, this is not to imply that one single "panacea product" has been developed. Obviously, a family of specifically tailored additives is necessary. Products for solvent-based, aqueous and solvent-free systems have been developed. Unique additives, which even make it possible to produce resin-free, aqueous pigment concentrates, have also been added to the product palette.

Six discrete advantages of the high molecular-weight polymerics are shown below:

- Universal binder compatibility
- Excellent dispersion and stabilization of the pigments
- No negative influence on weathering and/or durability
- Enhanced effectivity in high quality coating systems
- Improved pigment utilization
- Complete reproducibility of color and viscosity features

Question C: Exactly when is it possible to improve the deflocculation behavior of pigment "X" with a member of the high molecular-weight polymeric family? In other words, is there an easy-to-use, preliminary decision criterion which can help determine which additive is most appropriate?
The ascertainment of additive choice is generally quite simple; merely correlate the color index or pigment name with the appropriate additive name in Table 4.3.

To even further simplify the additive selection process, Appendices 1 and 2 have been specially prepared to display more than 200 starting formulations, to begin with, followed by additive usage levels for 250 different pigments (over 1800 system-specific additive combinations).

Table 4.3 [Please note that considerably less than 5% of the full cross-reference chart is displayed; only a representative excerpt (HMW #1) is included. Of course, several other HMW products exist and are therefore capable of covering a much wider pigment range than indicated in the chart below.]

	Colour index	Additive choice	Chemical pigment type
Yellow	P.Y. 83	HMW #1	Arylamid
	P.Y. 108	HMW #1	Anthrachinone
	P.Y. 110	HMW #1	Isoindolin derivative
	P.Y. 116	HMW #1	Arylamid
	P.Y. 139	HMW #1	Isoindolin derivative
	P.Y. 151	HMW #1	Arylamid
	P.Y. 154	HMW #1	Arylamid
Orange	P.O. 36	HMW #1	Arylamid (mono-azo-type)
Red	P.R. 88	HMW #1	Thioindigo
	P.R. 119	HMW #1	Naphthol AS
	P.R. 122	HMW #1	Quinacridone derivative
	P.R. 168	HMW #1	Anthanthrone derivative
	P.R. 170	HMW #1	Naphthol AS
	P.R. 177	HMW #1	Anthrachinone
	P.R. 178	HMW #1	Perylene
	P.R. 179	HMW #1	Perylene
	P.R. 224	HMW #1	Perylene
Violet	P.V. 19	HMW #1	Quinacridone
	P.V. 23	HMW #1	Dioxazine
Blue	P.B. 15	HMW #1	Phthalocyanine
	P.B. 60	HMW #1	Indanthrene
Green	P.GR. 36	HMW #1	Phthalocyanine
White	P.W. 6	HMW #1	Titanium dioxide
Red	P.R. 101	HMW #1	Iron oxide red
Black	(Special)	HMW #1	Channel and/or furnace black

*In selected instances, two or more alternate spellings of pigment names may exist. For the most part, "Americanized" spellings were employed; however, where exact American counterparts may not exist, European spellings were utilized.

Question D: After one has employed the pigment/additive cross-reference chart, what is the next step in the "decision analysis"?

The next step is to evaluate an initial test series, generally 5/10/15% (solids on pigment) additive; subsequent decisions are outlined in the decision-analysis flowchart displayed in Figure 4.3.

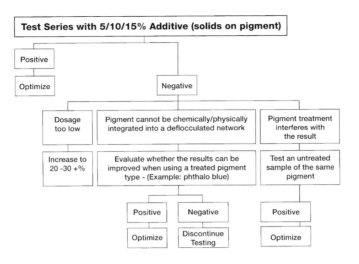

Figure 4.3 Decision-analysis flowchart

Question E: Why are relatively high dosages of HMWs sometimes necessary? Furthermore, from an economic perspective, does the magnitude of the resultant performance enhancement really warrant usage?

First of all, in response to the initial question, a synopsis of both the chemical and physical "working mechanisms" of HMW products is essential. As described below, HMW products function in a very unique fashion:

- The presence of up to a dozen or more specially designed "pigment adhesion groups" imparts radically improved wetting and dispersing behavior. ("Traditional" lower molecular-weight products, in comparison, generally contain only 1 to 4 pigment adhesion groups.)
- The determination of the exact role played by the aforementioned pigment adhesion groups is dependent upon many factors; nevertheless, the most crucial factor is the amount of pigment surface area available. Obviously, this is no surprise; the same rule applies whether or not one is employing HMW products. Organic pigments display higher surface areas and therefore require higher levels of wetting and dispersing agents. The decisive difference, however, with the higher molecular weight species (especially with organic pigments) is that the sheer abundance of pigment adhesion groups can allow (1) complete wetting of virtually all pigment surfaces, and (2) creation of a primary particle state (through electrostatic charge repulsion) in which an optimized, irreversible state of dispersion occurs. This means that, in contrast to traditional additives, HMW products are capable of "going a step further", so to speak. No pigment surface is left unwetted, and absolutely no interparticle contact or flocculation is allowed.
- High molecular-weight products, as mentioned above, employ electrostatic charge repulsion to create a primary particle state in which complete deflocculation occurs.

Exactly how is this achieved? Furthermore, what does this have to do with the necessity of relatively high usage levels in selected situations? The unique dispersion mechanism of HMW products allows the superimposing of positive charges over any and all previously existing charges in the coating system. Proper charge distribution therefore requires higher dosages of the wetting and dispersing additive. For instance, when one considers an actual case study with a challenging organic pigment (color index: P.R. 177), one finds that the pigment, as purchased, is negatively charged. (Please note that several hundred pigments have been tested, with the same basic trends and results being noted; the above case study was randomly selected.)

Now suppose that one were to inadvertently employ half the required amount of additive in conjunction with the aforementioned pigment. What would be the result? *Severe flocculation* would be the result since only half of the particles would receive positive charges. The second half would remain negative, and would subsequently be *attracted* to those pigments which had received positive charges. The end result, as one might predict, would be severe and irreversible flocculation. In field studies with the P.R. 177 pigment, such flocculation was exhibited at dosages of 7 to 9% additive. Nevertheless, merely doubling the additive level (as shown in the abbreviated graphic representation in Fig. 4.4) provided *absolute* deflocculation in the form of an empirically proven primary particle state. (Although a complete discussion of the empirical proof methodology is beyond the scope of this publication, a brief mention of its major points will be presented in answer to Question F).

The true utility of HMW products lies in their ability to achieve absolute control of particle size and distribution. From a practical standpoint, this means that one can achieve (as discussed in a subsequent section) statistically proven primary particle states.

Now—in direct response to the second part of the original question, a brief mention of the economic advantages offered by HMW products will be offered. Naturally, an

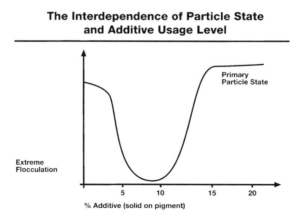

The Interdependence of Particle State and Additive Usage Level

Primary Particle State

Extreme Flocculation

% Additive (solid on pigment)

Figure 4.4 Particle state analysis

attractive economic incentive must exist; otherwise further discussion would be useless. Case studies with HMW products have demonstrated that rates of up to 60–70%* *less pigment utilization* (for equivalent color development) are not unusual. This economic advantage accrues in addition to the anticipated enhancements in universal binder compatibility and batch-to-batch reproducibility.

Question F: How can one empirically examine (with an easy-to-use methodology) primary-particle, controlled-flocculation, and/or uncontrolled flocculation states?
Of course, there are various techniques by which wetting and dispersing studies can be performed; this description overviews the methodology used to empirically study pigment particle states in situ. A completely functional, simplified version (without video attachments) of the experimental apparatus described below can be obtained for less than $50.00; by no means whatsoever is it imperative to employ all of the attachments mentioned.

- *Special Microscope Slide*
1. *Glass*: Standard transparent (at least 1"× 3"; but can be modified based on viewing apparatus).
2. *Copper strips*: Cat. No.: CF-3 DATAK Corporation or equivalent. (Generally, a printed circuit board grade is desirable. There are dozens of equivalent suppliers; local electronics shops may be consulted. Preferably, the strips should be 1/4" wide by at least 1" long. Distance between strips should be 0.7 to 1.0 mm.)

- *Microscope and Video*
1. *Microscope*: The recommended microscope (model "New Vanox T") can be obtained from Olympus and is equipped with an "F-10" television camera from Panasonic.
2. *Video*: For best results, one should utilize a professional S-VHS standard along with the following items or their equivalents: Camera BY-10E, Portable Recorder BR-S411E, Editing Recorder BR-S811E, Editing Control Unit RM-G410U. All of this equipment is available from JVC.

For computer-generated graphics, one can, of course, employ a wide variety of commercially available hardware and software combinations.

- *General Notes*
1. *Viewing*: Preparing ideal viewing conditions is often a reiterative process that requires several consecutive viscosity adjustments. Generally, lower viscosities provide mobility of a more visible nature. Propylene glycol methyl ether acetate has proved to be a successful diluting agent in some cases; however, different coating systems will, of course, require diluents of varying types.

---

*Of course, HMW products (like all coating ingredients) are very situation specific; obviously, no panaceas exist in all cases. Instances involving 60–70% pigment reductions are not commonplace with all pigments and with all formulations. In general, organic pigments (such as certain phthalo blues) prove to be more amenable to dramatic performance improvement shifts.

2. *Voltage/Amps/etc.*: Since this is not a "load-drawing" system, high amperage is neither preferred nor desired. Less than one amp is usually required; voltage should be controlled between at least the limits of 35 and 70V. A "variac" arrangement with fine control of milliamps and voltage is definitely recommended. Once again, the exact settings will be contingent on the actual test apparatus, the pigment system viewed, and the procedure employed.

Microscopic analysis of pigment wetting and dispersing phenomena is an exciting technology. In spite of all the idiosyncrasies involved with the set-up, much useful information can be gleaned from the process.

Question G: What might happen if one were to add too much HMW additive?
For all practical purposes, no deleterious effects (other than the extra raw material cost) would occur. High molecular-weight products do not negatively influence chemical and physical properties (such as corrosion resistance, durability, etc.).

Question H: Can pigment concentrates and/or full coating systems manufactured with HMW products be mixed with systems which do not contain HMW moieties?
Obviously, performance can be more easily optimized with uniform usage of HMW products "across the board", so to speak, but there are generally no systematic hindrances to employing both HMW and "non-HMW" products in the same mixture. As usual, this does not preclude the adherence to generally accepted wetting and dispersing principles. For instance, certain wetting and dispersing agent combinations (regardless of whether or not HMW entities are employed) could result in deleterious final coating properties. The well-known, classic examples of certain anionic/cationic combinations serve to illustrate this point quite well.

Question I: Can HMW products be "post-added"?
In principle, yes. Post-addition should ideally be limited to exceptional cases since optimal color development efficiency is achieved, by design, during the grind phase.

Question J: Suppose that one has no flocculation problems with his traditional "tried-and-true" grinding systems; should one still consider HMW products?
Given the wide variety of grinding systems encountered, one family of wetting and dispersing agents cannot serve as a performance-enhancing panacea in all situations; nevertheless—more often than not—incremental color strength increases can be achieved with HMW products. Even in problem-free, tried-and-true grinding systems, color strength improvements of 10 to 15% may be possible, especially in organic systems. Other possible advantages include absolute color and viscosity reproducibility along with improved processability.

Thus far, an integrated series of ten practical questions and answers has served as the springboard of our discussion; high molecular-weight additives provide the most optimal methodology for enhanced pigment wetting and dispersing performance. Several different HMW product families exist; therefore, formulators and manufacturers can choose from a wide palette of specially designed products. Scores of pigment/additive cross-reference tables have been made available with the express purpose of simplifying the additive selection process. One can also employ the inexpensive, easy-to-use "flocculation-state analyzer" described in this chapter. In conjunction with the important flocculation state factors discussed thus far, the three parameters outlined below are most crucial; in all cases, the achievement of optimal dispersion characteristics depends upon the mastering of *all three* parameters:

- Proper deflocculation state
- Homogeneously applied pigment charges (through utilization of HMW products where necessary and appropriate)
- Simultaneous attainment of both steric hindrance and uniform pigment mobility

Guide to Further Reading

Bode, R., Bühler, H., Ferch, H., Kautschuk und Gummi, *Kunststoffe* (1973), 26, p. 10

Coates, R., *Advances in Polymeric Wetting Additives for Solvent-Based Coatings* (1989) JOCCA 9

Degussa (Firmenschrift), *Prüfmethoden für Ruß* (1972) Frankfurt/M.

DIN: 53553 (1976), DIN: 53192 (1960)

Donnet, J.B., Lahaye, J., Voet, A., Prado, G., *Carbon* (1974) 12, p. 2

Elvira Moeller GmbH, *Druckfarben* (1980) Edition Lack und Chemie, Filderstadt

Endter, F., Anorg, Z., *Allgem. Chem.* (1950) 263, p. 191

Orr, E.W., *Principles of Pigment Wetting and Dispersing* (1995–1996) Lectures Available to FSCT Constituent Societies (special monograph series in support of FSCT lectures)

Schumacher, W., Ferch, H., Osswald, G., XII, *Fatipec-Kongreßbuch* (1974) p. 207

Schröder, J., *Morphology of Organic Pigments with Special Reference to Copper Phthalocyanine, Prog Organic Coatings* (1987), 12

Wagner, H., Schumacher, W., Ferch, H., XIII, *Fatipec-Kongreßbuch* (1976) p. 644

Wilsmer, L.C., *Pigm. Res. Techn.* (1972) 1, p. 30

5 The Real Difference Between Wetting and Dispersing

5.1 Conceptual Introduction

One of the most misunderstood concepts is the difference between wetting and dispersing; accordingly, this chapter sets forth definitions and describes the distinctive features of wetting "versus" dispersing—thereby dispelling the myths and misunderstandings that commonly surround pigment stabilization phenomena.

Since the incorporation of pigments into binder solutions is expensive and time consuming, a step-by-step analysis of performance improvement methods is presented. Special emphasis is placed upon defining the differences between wetting and dispersing. Both new theories and application data are systematically examined.

First, the function and mode of action of interfacially active substances are described. Next, the three parallel stages of the dispersion process are described in regard to effect, classification, and performance enhancement.

Pigment particles exist in one of three states or combinations thereof. In this discussion—flatting agents, fillers, and inerts are all called pigments. Primary pigment particles, as shown in the right-hand side of Figure 5.1, are normally what one is trying to achieve.

"Wetting and dispersing", as a term intended for usage in the coatings industry, should not be separated when discussing non-aqueous systems. This integrated term describes a process that first involves "wetting", a sub-process often characterized as the removal or displacement of air or moisture from the pigment surfaces, followed by the encapsulation of such surfaces with whatever is to be the continuous medium. The proper semantics of wetting and dispersing then involve an intermediate step, dispersing—the mechanical separation to primary particles—followed by the final step, the unequivocal stabilization of these particles to keep them from reflocculating. If the deflocculated system is not stabilized (symbolized by the necessity of "setting the parking brake" in the model car shown in Figure 5.2), then the system rolls back down

Aggregate   Agglomerate   Primary Particles

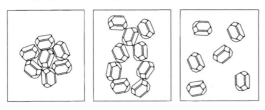

Figure 5.1 Comparative pigment states

Wetting and Dispersing Process

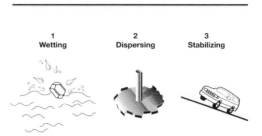

Figure 5.2 Three critical stages

the "energy hill", so to speak, thus reverting to a state of lower energy. Water-based and solvent-based systems utilize different additives due to their varying means of stabilization (such means of stabilization will be discussed in latter sections).

Differences in shade, sagging, flooding/floating, formation of Bénard cells, and settling are all problems associated with the wetting and dispersing of pigments. Wetting and dispersing additives accelerate the wetting of pigments in the binder and stabilize the dispersion of the pigments by steric hindrance or electrostatic repulsion.

Wetting and dispersing additives for solvent-based systems can be classified by their chemistry and their mode of action. Chemically, they are often classified as "anionic", "cationic" or "electroneutral". Experience shows, however, that this classification is not necessarily amenable to predicting the effectiveness levels of particular types.

A more meaningful classification is to differentiate between "deflocculating" and "controlled flocculating" modes of action; for instance, deflocculating types stabilize, to the greatest possible extent, the particles as single entities which are sterically separated from one another. That means a change in rheology toward Newtonian flow, thus:

- Excellent flow properties
- High gloss
- Best hiding power
- High transparency
- Development of precise color matches

Controlled flocculating types stabilize pigments and extenders to defined units of several particles; rheology is modified in a slightly thixotropic manner. Resulting advantages are:

- anti-sagging
- improved anti-flooding/floating

Figure 5.3 summarizes the features of not only controlled flocculation states, but also the other states that have been discussed so far.

As shown in Figure 5.4, in aqueous systems, the dispersing additive transfers electrical charges onto the pigment particle and stabilizes the system primarily via electrostatic repulsion. However, in organic systems, steric hindrance can often serve as the predominant means of stabilization. As described later, novel methods of charge

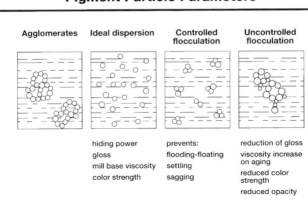

Comparative Dispersion States:
Pigment Particle Parameters

Agglomerates	Ideal dispersion	Controlled flocculation	Uncontrolled flocculation
	hiding power gloss mill base viscosity color strength	prevents: flooding-floating settling sagging	reduction of gloss viscosity increase on aging reduced color strength reduced opacity

Figure 5.3 Comparative features

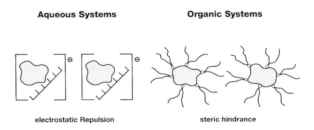

Aqueous Systems   **Organic Systems**

electrostatic Repulsion   steric hindrance

Figure 5.4 Aqueous versus organic systems

repulsion can also be used in solvent-based systems. (Special polymeric additives are required.) In water-based systems, the hydrophilic parts of an additive molecule (Fig. 5.5) can interact with the resin, thereby stabilizing the system.

In organic systems, the wetting and dispersing additive helps stabilize the pigment by steric hindrance. In such additives, at least one group is necessary which has a strong affinity to the pigment surface (anchor group), thus "adhering" the additive to the pigment particle. To effectively achieve steric hindrance, a protruding binder-compatible resin-like chain (Fig. 5.6) is necessary.

Typical structures for both deflocculating and controlled flocculating additives are shown in Figure 5.7. Note that the controlled flocculating product contains rather distinctively placed "pigment-bonding" groups that adhere not only to the pigment, but also to one another. (This feature helps control flood/float and sag.)

Ideal dispersion is more easily accomplished with stabilization when using the newer polymerics. These polymerics often improve gloss, color, hue, tint reproducibility, and film performance. Even though they are high molecular-weight products, they do not cause high viscosities. In fact, paint viscosity may even be reduced.

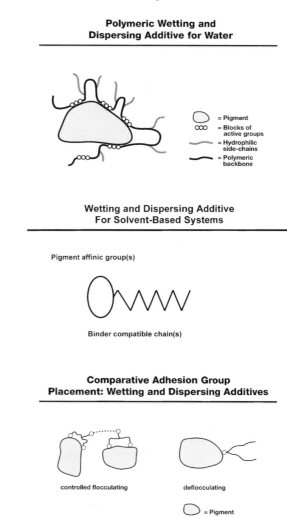

**Polymeric Wetting and
Dispersing Additive for Water**

= Pigment
= Blocks of active groups
= Hydrophilic side-chains
= Polymeric backbone

Figure 5.5 Molecular structure:
Polymeric additive

**Wetting and Dispersing Additive
For Solvent-Based Systems**

Pigment affinic group(s)

Figure 5.6 Elementary structural
components

Binder compatible chain(s)

**Comparative Adhesion Group
Placement: Wetting and Dispersing Additives**

controlled flocculating   deflocculating

= Pigment
= Adhesion group
= Polymer chain

Figure 5.7 Comparative placement

High molecular-weight polymerics stay more preferentially absorbed by the pigments, as opposed to establishing equilibrium, as do lower molecular-weight additives quite frequently. The polymerics stabilize by steric hindrance and by causing predictable electrostatic charge repulsion as shown in Figure 5.8. (A latter section will describe charge repulsion in more detail; a diagram of the charge measurement apparatus will also be shown.) Homogeneous positive charge repulsion aids stabilization and keeps pigment particles from reflocculating. The bottom half of Figure 5.9 shows proper stabilization.

Wetting difficulty is influenced by the surface area of the pigments being used. Mathematically, one can often estimate relative ease of wetting and dispersing with a

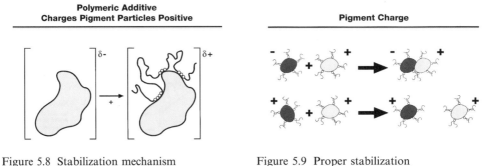

Figure 5.8 Stabilization mechanism

Figure 5.9 Proper stabilization

calculation of surface area and edge lengths. As shown in Figure 5.10, organic pigment particles may display a multi-thousandfold increase in composite surface area and edge-length statistics (compared to inorganic pigments), and may therefore be much more difficult to wet and disperse. Often, special polymeric additives with charge repulsion attributes and with many pigment adhesion groups are required.

5.2 Wetting versus Dispersing: What's the Difference?

Before one can adequately discuss a full-fledged program of performance enhancement and cost reduction methodologies, one must fully develop some of the introductory concepts mentioned in Chapter 3; accordingly, additional information is offered, in an expanded context, below.

Edge Length	Quantity of particles	Particle Surface	Total Sum of Edge Lengths
0.1 mm	1,000	60 mm^2	1.2 m
0.01 mm	1,000,000	6 cm^2	120 m
*1 μm	1,000,000,000	60 cm^2	12 km
+0.1 μm	1,000,000,000,000	600 cm^2	1200 km

* = inorganic + = organic

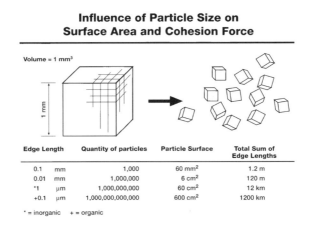

Figure 5.10 Geometric and mathematical relationships

5.2.1 The Wetting Phenomenon: Distinctive Features

Wetting additives reduce interfacial tension and, as a result, increase the "spreading pressure". Characteristic for such substances is their highly surface-active structure (polar, hydrophilic structural elements along with non-polar, hydrophobic structural elements combined in one molecule). Such substances migrate to the pigment/binder solution interface.

Influential factors include the polarities of the pigment surface and of the binder solution; the viscosity of the liquid phase; and also the geometry of the empty spaces (or pores) within the agglomerate structures. Of utmost importance is the interfacial tension in the exact areas where the wetting additive performs—between the pigment surfaces and the resin solution.

5.2.2 The Dispersing Phenomenon: Distinctive Features

In contrast to the previously described "wetting phase", the dispersing phase involves the *adsorption* of special additives onto the pigment surface to maintain proper pigment spacing through (1) electrostatic repulsion and/or (2) steric hindrance, thus reducing the tendency to uncontrolled flocculation. Unique aspects of both pigment spacing techniques are overviewed below:

5.2.2.1 Electrostatic Repulsion

The pigment particles in the liquid paint carry electrical charges on their surfaces. Through the usage of special additives, it is possible to strengthen the charges and, furthermore, *to make all pigment particles equivalently charged.* Counterions concentrate around the pigment particles so that an "electrical double-layer" is formed; stabilization increases along with layer thickness. This particular electrostatic repulsion stabilization mechanism is especially useful in water-based latex dispersion systems. Chemically speaking, the additives utilized for dispersion in such systems are polyelectrolytes—tailored higher molecular-weight products with electrical charges in the side chains. In addition to polyphosphates, many polycarboxylic acid derivatives are employed as polyelectrolytes in the coatings industry. The polyelectrolytes adsorb onto the pigment surface and consequently transfer their charge to the pigment particle. Through electrostatic repulsion between equally charged pigments, the deflocculated state is stabilized.

5.2.2.2 Steric Hindrance

Dispersing additives functioning by steric hindrance display two special structural features. First, such products contain one or more so-called "pigment-affinic" groups—anchor groups or adhesion groups—all of which provide strong adsorption upon the pigment surface. Second, such products contain resin-compatible chains (hydrocarbon entities) which, after additive adsorption upon the pigment, protrude as far as possible from the pigment surface into the surrounding resin solution. This layer of adsorbed additive molecules with the protruding chains produces steric hindrance or "*entropic stabilization*".

The above entropic stabilization is further accentuated by the interaction of the additive's polymeric segments with the resin polymers in such a way that the "envelope", so to speak, around the pigment particles is enlarged. This stabilization mechanism occurs in solvent-based systems and also in certain water-reducible systems. Through specific structural elements composed of pigment-affinic groups (polar) and resin-compatible chains (non-polar), the polymeric additives exhibit definitive surface-active properties. In other words, they not only stabilize the pigment dispersion, but they also function as wetting products.

The concept of "flocculation" obviously carries rather negative connotations; however, there are indeed certain conditions under which *controlled* flocculation may actually be more desirable than complete deflocculation. Subsequent discussions will describe (1) how additives can be selectively designed to provide deflocculating or controlled flocculating effects, and (2) the particular conditions under which one stabilization mechanism or the other is more desirable or appropriate.

5.3 Deflocculation: How and Why?

The structure of classical deflocculating additives has already been described as containing one or more spatially close pigment affinic groups within the molecular framework of a number of resin-like chain structures. Classical additives contain low molecular-weight polymers that adsorb upon the pigment surface and that stabilize the deflocculated condition by steric hindrance. Deflocculation generates a rather Newtonian flow behavior along with reduced viscosity. In this manner, flow behavior is enhanced and a higher pigment level is possible. Due to the small particle size of the deflocculated pigments, both increased gloss and optimized color strength are obtained.

Deflocculation generally leads to an improved, more efficient pigment utilization, which (especially in the case of the sometimes rather expensive organic pigments) is not economically unimportant. The degree of deflocculation or flocculation exerts a dramatic influence upon the developed shade or tint of a pigment. If, for example, a system tends to settle upon storage, then color shift can result. In situations where this is especially critical (such as in the base component of a mixing system), the only acceptable method for producing coatings with a constant and defined shade is the *complete deflocculation* method achieved by high molecular-weight additives.

5.4 Controlled Flocculation: How and Why?

If the pigment affinic groups are not merely confined to a small region of the additive molecule, but are distributed in a special fashion over the entire molecule—then they can simultaneously contact two or more pigment molecules in a "bridge-like" fashion. Controlled flocculation is the result. At this point, it is important to clarify the difference between the above controlled flocculation state and the "normal" flocculation state. Without additives, the pigment particles make direct contact with one another in uncontrolled flocculates. In contrast, no direct pigment/pigment contact occurs in controlled flocculates; additive molecules are always present between the pigment particles.

Although uncontrolled flocculation is always undesirable (since a variety of negative side-effects may occur), *controlled flocculation* may purposefully be employed to attain certain desired features in the coating system. The controlled flocculation state normally forms three-dimensional wetting structures which lead to thixotropic flow behavior within the coating. Through these structures, the resting state viscosity is rather high; however, when shear forces are applied, the structures (pigment flocculates) break apart and induce lower viscosity. Afterwards (following removal of the shear forces), the flocculates can rebuild. Generally, such systems display a more or less well-pronounced flow threshold. By utilizing the above rheological behavior, deleterious properties such as sagging and settling can be alleviated. Through controlled flocculation, both flooding and floating can be eliminated since the different pigments are bound together in the flocculates (and therefore cannot separate from the mixture). The building of flocculates can, under certain conditions, naturally reduce coating gloss; systems should be evaluated on a case-by-case basis for possible undesirable side-effects.

Especially in primers and primer surfacers, lower gloss is generally acceptable. Controlled flocculating additives can, at times, be utilized in topcoat systems; dependent upon the resin system and the additive usage level, lower gloss may not necessarily occur. The primary application area for controlled flocculating additives is found in protective coating systems, whereas deflocculating additives demonstrate special utility in systems demanding optimal appearance and optical properties. Controlled flocculating additives are often used in combination with other rheological additives (fumed silicas, dehydrated castor oils, and bentonites) because of frequently occurring synergistic effects.

5.5 Special Considerations in Aqueous Systems

Electrostatic repulsion and steric hindrance are the stabilizing mechanisms for water-based systems. The question of exactly which mechanism is more appropriate depends primarily upon the resin choice. Water-soluble resins are composed of water-solvated polymer chains; accordingly, they are physically very similar to the solvent-based resin systems. The solvated, unfurled resin chains can reinforce steric hindrance. This explains why the aforementioned stabilization mechanism is so important. The same principle is applicable for hybrid systems (mixtures of water-soluble resins with emulsions) since such systems also contain resin portions which are solvated. In order for additives to be capable of producing steric hindrance in such systems, it is essential that the additives' polymeric segments be compatible with the aqueous binder solution. It is, perhaps in a rather surprising manner, not necessarily desirable for the additive to be water-soluble, since durability could be adversely affected. A proper balance must be achieved. One should ideally transfer the experience gained from formulating solvent-based systems to the aqueous coatings field; this means that both classical, low molecular-weight polymers as well as the more modern, high molecular-weight polymeric additives can be employed. *In many instances, wetting and dispersing additive technology has advanced to the point where even resin-free pigment concentrates (containing only pigment, additive, and water) can be produced.*

The high dielectric constant of water makes strong stabilization through electrostatic repulsion possible. When one considers pure emulsion systems, then electrostatic repulsion is generally the major stabilization mechanism since no solvated and bonded polymer chains are available for proper steric hindrance. (In contrast, water-soluble and/or hybrid systems are not as dependent upon the electrostatic repulsion stabilization method.) Dispersing additives which strengthen electrostatic repulsion are often salts of polycarboxylic acids. The degree of stabilization depends upon the thickness of the electrical double layer; and this layer's thickness is, in turn, strongly influenced by contaminating ions and electrolytes originating from a variety of sources, including from pigments and fillers. The level of electrostatic repulsion is accordingly reduced as the charge and concentration of such contaminating electrolytes increase. On the other hand, steric hindrance is simply not affected by such electrolytes. In the latter stabilization method, the influence of the organic co-solvent is more important, especially considering that most water-based formulations contain small quantities of co-solvents. These organic co-solvents can alter the solvation properties of the dissolved polymeric structures and can thus bring about considerable viscosity influences.

Guide to Further Reading

Degussa AG, Schriftenreihe Pigmente Nr. 16, *Untersuchungsmethoden für synthetische Kieselsäuren und Silikate* (1977) Frankfurt/M., 3. Auflage

DIN 53192 (1960), DIN 53234 (1972)

Medalia, A.I., Eaton, E.R., Kautschuk und Gummi, *Kunststoffe* (1967) 20, p. 61

Mildenstein, E., *Die Komponenten des Dispergiergutes, ihre Wechselwirkungen und die Optimierung der Zusammensetzung. Ingenieurarbeit SS* (1972) FH Niederrhein, Abt. Krefeld, Fachbereich Chemie (Lack und Kunststoff)

Orr, E.W, *How to Enhance Performance with Additives*, lectures and monographs presented at DePaul University (1995)

Orr, E.W., *Performance Improvement with Additives*, lectures and monographs presented at Eastern Michigan University, (1994–1995)

Rehacek, K., Untersuchung der Wechselwirkung zwischen Pigment und Bindemittel in konzentrierten Bindemittellösungen, *Farbe + Lack* (1979) 76, p. 656–665

Weigl, J., *Elektrokinetische Grenzflächenvorgänge* (1977) Verlag Chemie, Weinheim

6 The Control of Flooding and Floating: Theory and Practice

6.1 Introduction

The problem of flooding and floating in pigment combinations is so complex that only a brief summary of the most important factors can be provided.

Definition: Even though various segments of the industry may employ several different situation-specific definitions, "flooding and floating" always involves a color change in the still-liquid coating. This fascinating and controversial phenomenon is caused, in part, by a differential concentration of one pigment species from the pigment mix; furthermore, one may choose to demarcate fine lines of distinction between "flooding" and "floating" as follows:

- *Flooding* is horizontal separation
- *Floating* is vertical separation

Causes: The typical causes of flooding and floating are shown below:

- Flocculation of one of the pigments in the mix
- Differences in pigment mobility
- Currents in the coating

6.2 The Physicochemical Background

6.2.1 Flocculation of one of the Pigments

Flocculation is the agglomeration of the primary particles to form a loose agglomerate which, in flooding and floating, then behaves as a pigment with a large diameter. Flocculation means a loss in tinting strength for the flocculated pigment. One must differentiate between three forms of flocculation:

- Cohesive flocculation: The individual pigment particles touch each other and cohere, as a result of their own surface forces.
- Pigment/binding agent flocculation: The pigment and binding agent flocculate together.
- Bridging flocculation: The pigment particles agglomerate with other substances such as water, stabilizers, and wetting agents.

On one hand, flocculation depends on the pigment itself, and on the other hand, on the coating system in which it appears. Wetting agents and the various methods of pigment dispersion also exert considerable influence on flocculation. Organic pigments of small particle size and large surface area tend to flocculate more than most inorganic pigments, which have larger diameter, but smaller surface area. As a result of their diminutive size, the smaller pigment particles move significantly faster, collide more frequently, and thus have more opportunity to encounter other surfaces, and so, to flocculate. Experience confirms that organic pigments are more subject to flocculation. In every coating formulation there are chemical compounds which could attach themselves to the surface of the pigment. Examples include stabilizers, slip agents, resins, etc. To prevent these uncontrolled attachments and the resultant reactions, it is important to treat the pigments, during dispersion, with special wetting and dispersing aids.

6.2.2 Differences in Pigment Mobility

Differential pigment mobilities are caused, in part, by the following factors:

- Particle size
- Degree of flocculation
- Specific weight
- Electric charge

The main cause is often particle size. In turn, particle size is integrally related to the degree of flocculation, especially since agglomerates behave as large pigment particles. As shown in Figure 6.1, relative pigment mobilities can vary dramatically—from 2 to 100,000. (The author's laboratories, along with the research staffs of S. H. Bell and others, have corroborated the existence of the aforementioned mobility differences.) With differences of this magnitude, it is understandable that pigment mixtures may separate in the currents which form during the drying process.

6.2.3 Currents

Solvents evaporating from low viscosity and solvent-containing coatings cause eddy currents, which, according to the Helmholtz flow distribution theory, form an intense

Pigment	Relative Mobility
• Titanium dioxide	2-10
• Iron oxide yellow	25
• Prussian blue	500
• Carbon black	100,000

Figure 6.1 Relative pigment mobilities

network of irregular hexagons known as Bénard cells. The pigments are caught up in the eddy currents, and their differing mobilities result in separation and floating.

Pigment movement during flooding and floating depends on the pigment diameter. Since the rate of fall of a pigment is opposed by the upward movement of flooding and floating, experience has shown that large-diameter pigment particles (which, according to Stokes' Law, have a greater rate of fall than small-diameter pigments) gather on the surface to a markedly lesser extent than their small-diameter counterparts.

Although one permutation of Stokes' Law (Fig. 6.2) applies to the rate of fall of spherical particles in liquids, various factors (such as differences in wetting of the pigments, nature of the pigment surface, pigment-volume concentration, etc.) prohibit an exact application.

Regarding the influence of particle size on floating, a simplified formula based on Stokes' Law is useful; since a constant medium can be assumed, a simplified expression for the rate of fall of pigment particles can be derived: $V = (d_{Pi} - d_{PL})r^2$. By comparison of the values derived by the use of this expression, it is possible to predict the flotation tendency, provided the pigments do not flocculate.

One may illustrate the rate of pigment fall with a coating containing titanium dioxide and red iron oxide; the corresponding mathematical relationships are shown in Figure 6.3. As shown, the rate of descent of the TiO_2 is demonstrated to be approximately 20 times greater than that of the Fe_2O_3, even though the specific gravity of the Fe_2O_3 is higher. Since the rate of fall is opposed by the upward motion of flotation, the red iron oxide will, in this example, accumulate on the surface.

If one examines a typical light blue coating (containing TiO_2 and phthalocyanine blue) as an example, the calculation shows the rate of TiO_2 fall to be approximately 80 times that of the blue pigment. (Please note that only the end results of this particular

$$V = \frac{D}{\eta}(d_{Pi} - d_{PL})r^2$$

Definitions

V	= Rate of fall
D	= Proportionality constant
η	= Viscosity of coating
d_{Pi}	= Density of pigment
d_{PL}	= Density of coating
r	= Radius of pigment particle

Figure 6.2 Rate of fall

Fe_2O_3: $d_{Pi} = 5.0$ g/cm^3 $r = 0.05\mu m$

TiO_2: $d_{Pi} = 4.1$ g/cm^3 $r = 0.25\mu m$

$d_{PL} = 1.2$ g/cm^3

V' $Fe_2O_3 = (5.0 - 1.2) \cdot 0.05^2 = 0.0095$

V' $TiO_2 = (4.1 - 1.2) \cdot 0.25^2 = 0.18$

Figure 6.3 Differential rates of fall

case study are presented; the intervening calculations are not shown.) However, the coating, upon close examination, demonstrates that both pigments are flooding and floating. Microphotographs show that the phthalocyanine blue flocculates to such an extent that the increase in the radius of the blue pigment particles explains, in part, the flooding and floating of the TiO_2. The remaining unflocculated phthalocyanine blue also floods and floats, as the values derived from the germane mathematical formulas indicate.

6.3 Prevention and Control Mechanisms

Pigment mixtures will exhibit floating if the various pigments exhibit different mobilities. One method of eliminating the differences in pigment particle mobility is to co-flocculate the pigments. By using dispersing agents which produce controlled flocculation, the different pigments are jointly incorporated into flocculates so that they can no longer move independently. This process moves the paint further away from the desirable state of complete deflocculation; nevertheless, this method can often be successfully used to avoid flooding and floating.

6.3.1 Prevention of Floating through Deflocculation

As discussed previously, complete deflocculation of all pigments is absolutely essential if maximum gloss and color stability are required. If, however, all the pigments are deflocculated, the differences in particle size between organic and inorganic pigments—often a factor of 10 or more—will be very obvious, as will be the differences in mobility. The question, therefore, is how one can also prevent floating in deflocculated paint systems.

One solution is to use high molecular-weight dispersing agents whose protruding polymer chains can strongly interact with the surrounding binder molecules, so that the mobility of the pigment particles is greatly reduced. Practical experience has shown that the effect of differences in size between organic and inorganic pigments can be eliminated in this fashion. Accordingly, pigments of various sizes can be tailored to display similar mobilities.

Since polymeric dispersing agents show excellent adsorbency on organic pigments, they can be used to suppress floating in pigment blends. At the same time, the pigments are deflocculated, which is an important factor for high quality topcoats. Comparison of different polymeric dispersing agents shows that, although they all deflocculate organic pigments (evident, for example, in increased pigment transparency), one of the most efficient products for preventing floating in mixtures with a white base is Polymer Additive A. (Please see appendices for additional information. Generic compositions and specifications, at least for a representative cross-section of many additives mentioned, are contained at the end of Appendix I. Virtually no additive trade names are mentioned; of course, there exist several suppliers for all additives mentioned in this text.)

6.3.2 The Influence of Electrical Charges: A Case Study (with Performance Flowchart)

In waterborne coatings, the electrostatic repulsion of pigment particles carrying the same charge forms the basic mechanism for stabilizing the deflocculated state. Ionic charge values are generally weaker, though, in paints containing organic solvents. As a result, deflocculation is often achieved in solvent-based systems through steric hindrance. Nevertheless, relatively weak charges can indeed dramatically affect the stability of pigment mixtures, regardless of whether the systems are water- or solvent-based. The charge carried by a pigment can be easily determined, at least qualitatively, with a flocculation-state analyzer similar to the device mentioned in a previous chapter. Two copper strips (Fig. 6.4) are arranged at a distance of about 1 mm from each other on a glass slide, forming the electrodes. The slide is placed under a microscope and a drop of the greatly diluted paint is placed between the electrodes. After application of a DC current of approximately 60 V, the migration of the pigment particles in the electrical field can be observed. The direction of movement reveals their charge—positive or negative.

The type of charge carried by a pigment depends on the pigment itself as well as on the binder solution (Tables 6.1 and 6.2); accordingly, the following rules apply:

- A given pigment can carry different charges in different resins
- Different pigments ground in the same resin can carry different charges

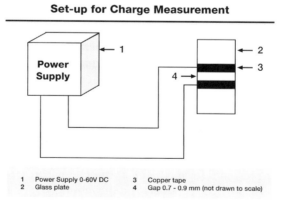

Set-up for Charge Measurement

| 1 | Power Supply 0-60V DC | 3 | Copper tape |
| 2 | Glass plate | 4 | Gap 0.7 - 0.9 mm (not drawn to scale) |

Figure 6.4 How to measure pigment charge

Table 6.1   Different Resins Exhibiting Different Charges with the Same Pigment

Binder	Charge
Macrynal SM 515	$++$
Setalux C 1502	$--$
Alftalat A451	0

Explanation	
$+, ++, +++$	= weak, strong, very strong positive charge
$-, --, ---$	= weak, strong, very strong negative charge
0	= no detectable charge (using method described in text)

Table 6.2   Different Pigments Exhibiting Different Charges with the Same Resin

Pigment	Charge
Bayferrox 130 m	$+$*
Special black IV	$---$
Cinquasia red Y FT 859 D	$++$
Chromophtal red A2B	$--$

*For explanation of symbols, please see Table 6.1. Please note that, as described in Table 4.3, there may exist alternate European and American spellings of certain pigment names.

Table 6.3   Provision of Positive Charge with an Additive

Pigment	Pigment grind	Charge
Bayferrox 130 m	Reference sample	$+$*
	Polymer additive I	$++$
	Polymer additive II	$+$
Special black IV	Reference sample	$--$
	Polymer additive I	$++$
	Polymer additive II	$-$
Cinquasia red Y RT 859 D	Reference sample	$++$
	Polymer additive I	$++$
	Polymer additive II	$++$
Chromophtal red A2B	Reference sample	$--$
	Polymer additive I	$+$
	Polymer additive II	0

*For explanation of symbols, please see Table 6.1.

The fact that the pigment charge can be affected by the deposition of additives on the pigment surface is of practical importance. All pigments can be given a positive charge by using an appropriate dispersing agent. This is shown in Table 6.3 for various pigments. In addition, Table 6.4 shows how to incorporate such polymeric additives

Table 6.4   Flowchart: How to Prepare Optimal Pigment Concentrates

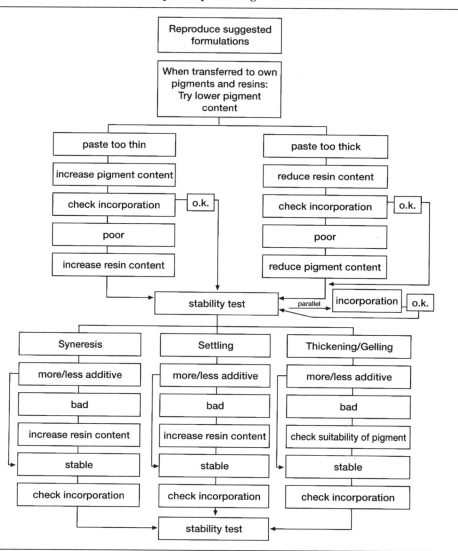

into the preparation of optimal pigment pastes. Not only can degree of flocculation be measured with the charge device described previously, but degree of flocculation can also be roughly estimated from a color shift diagram such as that shown in Table 6.5.

In conclusion, an appropriate dispersing agent can indeed be quite powerful. As displayed in Table 6.3, all pigments were given a positive charge when the proper polymeric additive was employed. Coatings manufactured with optimally dispersed pigments will, of course, display superior overall performance.

Table 6.5 Chromaticity Diagram in Accordance with CIE Publication 15.2, ASTM E 308 and DIN 5033

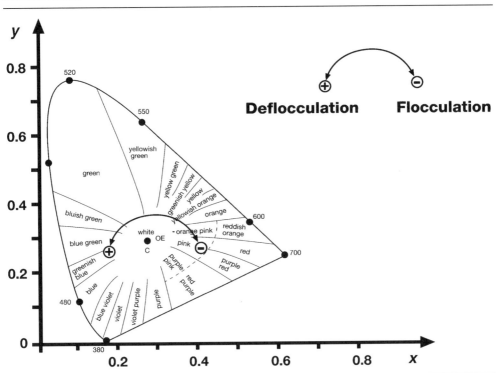

Guide to Further Reading

Bell, S.H., *J. Oil Colour Chem. Assoc.* (1952) 35, p. 386

Degussa Corporate Communication, Prüfmethoden für Russ (1972) Frankfurt/M

Gall, L., Kaluza, U., *Grundlagen der Pigmentdispergierung*, Defazet (1975) 29 3, p. 102–106

Gengenbach, O., Schene, H., Maisch, W., *Die elektrophoretische Beschichtung von Metallteilen, ein neues Lackierverfahren*, Mitteilungen der Forschungsgesellschaft Blochverarbeitung (1965) p. 14

Green, H., *Industrial Rheology and Rheological Structures* (1949) John Wiley and Sons, Inc., New York

Holliday, L. (Ed.) *Ionic Polymers* (1975) John Wiley and Sons, Inc., New York

Kresse, P., *Farbe + Lack* (1966) p. 111

Lubs, H.A. (Ed.) *The Chemistry of Synthetic Dyes and Pigments* (ACS Monograph) (1965) Hafner Publishing Co., New York

Orr, E.W., *The Control of Flooding and Floating*, Lectures Available to FSCT Constituent Societies (special monograph series in support of FSCT lectures) (1995–1996)

7 How to Control Gloss, Haze, and Color

7.1 Introduction

Gloss, haze, and color are influenced by a large number of variables. In addition to the basic coating formulation itself, the following specific parameters are vitally important:

- Pigment properties—dispersion behavior and dispersibility
- Compatibility of binders and additives
- Mill base formulation
- Dispersion method
- Application method
- Substrate
- Drying conditions

The first four parameters above are of particular importance since gloss and reflection haze often correlate with particle size in the final film. Coarse particles—whether primary particles, aggregates or flocculates—cause surface imperfections which result in scattered light and gloss reduction.

7.1.1 The Influence of Dispersion State on Gloss, Haze, and Color

Poorly dispersed pigments or flocculates cause microscopic surface defects which result in scattered light of low intensity adjacent to the direction of specular reflection. When visually observed, this scattered light causes a milky appearance called *reflection haze,* or simply *haze*.

Selected case studies will be discussed (in subsequent sections of this chapter) with the intent of illustrating the meaning of gloss and haze measurements for evaluating the degree of dispersion. In addition, several germane factors, especially those which critically influence dispersion results, will be examined.

7.2 Coating Production, Sample Preparation, and Measurement

Dispersion was performed with dispermat laboratory dissolvers. Thinned-down paints (approximately 22 to 25 s; DIN 4 mm; determination of viscosity by means of flow cup) were poured onto glass plates for the studies performed in Section 7.4 of this chapter. The pouring method is particularly suitable for "ladder-studies" and reiterative series of measurements. Since this application method avoids shearing, then any flocculation—even if extremely small—will be detected. The result is a thin-paint film that allows excellent differentiation of performance properties.

A polished glass substrate was used expressly to avoid defects caused by a rough or uneven substrate, which would have made both the visual evaluation and the physical testing considerably more difficult. To facilitate reproducibility, the mean readings of 3 to 5 measurements were calculated.

7.3 Wetting and Dispersing of Pigments: Special Considerations

Before the case studies in Section 7.4 can be properly introduced, it is essential to briefly mention selected background information in the production measurement and evaluation arena. One of the most important steps in coatings production is the fine and uniform distribution of the solid pigment (and/or filler) particles within the resin solution. This procedure can undoubtedly be very time consuming and cost intensive, particularly if a high-gloss paint is desired. A dispersing process that does not take place optimally may cause the following undesired effects:

- Flocculation
- Loss of gloss
- Reflection haze
- Floating
- Bénard cells
- Sedimentation

When producing high-gloss paints, the degree of dispersion is an important concern; accordingly, ISO 8780 P 1-6 and ISO 8781 P3 standards (among others) describe the testing of dispersing behavior. (Please note that ISO/DIN/ASTM and/or other standard numbering and cross-reference systems are often modified, sometimes without notice, by their respective promulgating committees. Certain alphanumeric designations are even changed on a quarterly basis; therefore, the reader is encouraged to consult the standard quarterly periodicals.) For evaluating performance in a statistically representative fashion, the degree of dispersion is usually measured at different stages and/or by incorporating a variety of production-specific variables (i.e., dispersion time, dispersion passes, or number of revolutions) into the test battery. As an integral part of the evaluation program, the following two measuring methods may be employed:

- Direct measurement of particle size
- Microscopic evaluation (sometimes complicated, may be unsuitable for certain types of process control)

Measurements taken with a grindometer per DIN 53 203 and ASTM D 1210 provide useful information about poorly dispersed agglomerates. Such measurements also constitute a fast and easy method to evaluate the dispersion process. The fineness of grind will improve with the degree of dispersion; for instance, the number of coarse particles decreases and the number of fine particles increases. Generally speaking, the limits of these methods are reached if the pigments contain particles and agglomerates predominantly smaller than 1 μm.

7.4 Dispersing Behavior

7.4.1 Inorganic Pigments

In a test study, the dispersing behavior of TiO_2 was evaluated in a representative system with an inorganic pigment. Degree of dispersion was evaluated after dispersion times of 5, 15, 25, and 45 min. A direct particle size determination with a grindometer was undertaken; in addition, gloss and haze measurements were performed. For simplicity's sake, all measurements were expressed in a relativistic (0 to100) fashion.

The following baking enamel was tested:

Binder(s):
Alkyd resin
Melamine resin
Pigmentation:
Titanium oxide (Pigment White 6)
Additive:
High molecular-weight anionic additive

Based on the test results, one can conclude that a reduction in fineness of grind does not always cause a considerable change in gloss, but it can indeed lead to a significant reduction of the reflection haze value (Fig. 7.1). This demonstrates that, in certain system-dependent ranges, reflection haze measurement is a particularly valuable tool

Fineness of Grind, Gloss (R'20°) and Haze Related to the Dispersion Time (inorganic-pigmented)

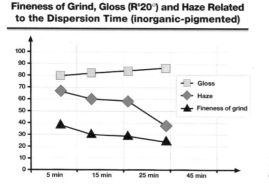

Figure 7.1 Three variables as related to dispersion time

for dispersion evaluation, whereas fineness-of-grind determination by means of grindometer is comparatively limited.

7.4.2 Organic Pigments

For evaluating the dispersion of organic pigments, fineness-of-grind determination with a grindometer is simply not sufficient. Indirect measuring methods are required. For this reason, dispersibility is often determined by measuring the tinting strength development, the latter being defined as a measure of the colorant's ability to add color to other materials (based on absorbing capacity as specified by DIN 53 243). For the express purpose of determining tinting strength, the organic-pigmented paint is dispersed for different time periods, and is then subsequently mixed with an appropriate white enamel. A spectrophotometer is used to determine the K/S-value (derived from a modified Kubelka–Munk equation; representative K/S values are displayed in Figure 7.2) in the absorption maximum; accordingly, a plot of this value versus time can be shown. The K/S value is a practical measure of the tinting strength. The better the degree of pigment dispersion, the greater the tinting strength. This method does obviously include, however, many possible sources of error, such as those involved with weighing procedures or segregating tendencies. Nevertheless, these sources of error can be rather effectively eliminated, though, with gloss and haze measurements–since neither weighing nor mixing is required. Measurements are simply taken from the "masstone" of the color. In the following laboratory study, a baking enamel was tested:

System:
Alkyd resin (synthetic fatty acid) / melamine resin
Pigmentation:
Organic red pigment (C.I. Pigment Red 254), titanium dioxide (C.I. Pigment White 6) for the white admixture. (Please note that many C.I. designations reflect supplier-modified interpretations of C.I. nomenclature.)
Additive:
High molecular-weight cationic additive

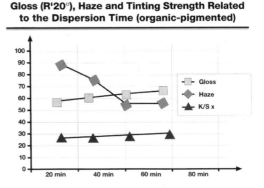

Figure 7.2 The superiority of haze as a method to evaluate dispersion quality

In the relativistic example shown in Figure 7.2, haze measurement proved to be the optimal means of evaluating dispersion quality. Neither gloss nor tinting strength allowed a clear and unequivocal evaluation.

7.5 Influence of the Binder System

The interaction between binder and pigment is dependent upon the disparate chemical natures of the ingredients encountered, and can obviously change rather drastically; therefore, the binder's influence on pigment dispersing behavior and, as a consequence, on the reflection properties, is a crucial factor worthy of study.

In the following example, two baking enamels were prepared using the same pigment and the same PVC, but with different resins.

Pigmentation:
Organic red pigment (C.I. Pigment Red 254)
Additive:
High molecular-weight cationic additive
Binder A:
Alkyd resin (synthetic fatty acid) / melamine resin
Binder B:
Saturated polyester / melamine resin

In this study, reflection haze measurement (Fig. 7.3) resulted in clear differentiation between the two samples. The visually observed difference between samples A and B was not at all demonstrable with gloss measurement, but was quite evident with haze measurement.

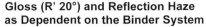

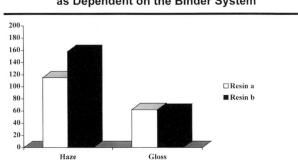

Figure 7.3 Further confirmation of haze as a superior differentiating factor

7.6 Wetting and Dispersing Additives

By definition, wetting and dispersing additives accelerate the wetting process in the binder, and furthermore stabilize pigment dispersion by steric hindrance and/or electrostatic repulsion. Depending upon one's usage and application area, most pigments are stabilized as individualized particles (as opposed to specially interlinked particles in a controlled flocculation state) and are sterically separated. This means a change in rheology towards Newtonian flow behavior; namely, the following advantages accrue:

- Improved flow
- High gloss
- No reflection haze
- Optimal hiding power
- High transparency
- Development of pure chroma

7.6.1 Optimization of Optical Properties

The optical properties of coatings can be positively influenced by the use of wetting and dispersing additives. For instance, to obtain a glossy appearance, deflocculating additives are normally used; nevertheless, how can one properly gauge the resultant optical enhancement—assuming that one's choice of gloss-improving additive was appropriate? As already shown in Figure 7.3, tandem gloss and reflection haze measurements are useful initial tools; however, one can even further accentuate the measurement possibilities by, for instance, subdividing haze measurements into the two categories shown in Figure 7.4. (Please note that, because of space limitations, gloss measurements are not shown in Figure 7.4—especially since they resulted in little or no differentiation of performance.) Dual haze measurements in the following baking enamel were evaluated, with the end result being a series of rather dramatic

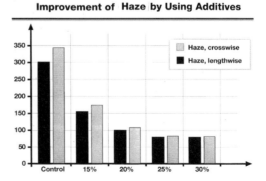

Figure 7.4 The role of wetting and dispersing additives

haze improvements (meaning that the measured values decreased) with increasing additive levels:

System:
Alkyd resin/ melamine resin
Pigmentation:
Red iron oxide
Additives:
Polymeric Additive "J"
Polymeric Additive "M"

7.6.2 Additive Concentration

To determine the optimum additive concentration, it is recommended that one always perform systematic ladder-studies and evaluation programs with differing amounts of a wide variety of candidate additives. In addition to the still-necessary color and transparency measurements, gloss and reflection haze are excellent evaluation tools in most systems. For instance, the final phase in the optimization of additive concentration for a typical polyester system is shown in the following example:

System:
Saturated polyester resin/melamine resin
Pigmentation:
Organic red pigment (C.I. Pigment Red 254)
Additive:
Polymeric Additive "C" in different concentrations (all references here are expressed in percentage of active substance on pigment)

Figure 7.5 demonstrates that the ideal performance mix—optimal gloss improvement and, at the same time, dramatic reflection haze reduction—are achieved when adding up to 25% of the additive.

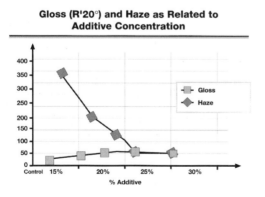

Figure 7.5 Gloss improvement and haze reduction

7.6.3 Influence of the Application Method

To obtain the most informative and reproducible measurement results, it is necessary to examine the influence of different application methods on gloss and reflection haze. For many series of laboratory tests, pouring the paint onto glass proves to be—at least within the context of performance differentiation—a suitable application method. Often, however, a film applicator may be more appropriate. Many laboratories include, and rightly so, the application method actually used in large-scale production (e.g., in industrial processes—multitudes of spray systems, bell applicator devices, dipping baths, curtain-coater lines, tailored UV/EB/powder applicators, hybrid scenarios, and/or special deposition methodologies, etc.). Three different application methods (pouring, film applicator, and spraying) are compared in Figure 7.6, where a one-coat paint system was applied. In the development stage, this paint was first tested with "pour-outs". As evident from the bar chart, the *poured* samples exhibited lower gloss values and higher reflection haze values, both of which are definite indicators of flocculation. In contrast, though, when using a film applicator, or particularly when spraying, the flocculates were often destroyed by shear forces. On a comparative basis, this led, therefore, to higher gloss and lower reflection haze values.

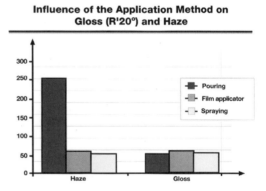

Figure 7.6 Differential application methods

For the tailored evaluation of either the dispersion state or the colloidal stability of paint, the pouring methodology—particularly when used in combination with reflection haze evaluation—has definitely proved its merits. The overt presence of flocculates, or even the very hint of their presence, can often be detected quite easily with this methodology.

Guide to Further Reading

Berrie, P.G., *Schichtdickenmessung nach dem coulometrischen Verfahren, universell, flexibel und wirtschaftlich* (1990) Fischer GmbH Statistik in der analytischen Chemie (1990) VEB Deutscher Verlag für Grundstoffindustrie, Leipzig, S. 50, S. 90

Eisenmenger, E., *Pitture e Vernici* (1973) 49, p. 311 Chim. D. Peintures (1973) 36, p. 254

Ferch, H., FATIPEC-Kongreßbuch (1966) Verlag Chemie, Weinheim

Hollemann-Wiberg, *Lehrbuch der anorganischen Chemie* (1985) Walter de Gruyter Seifen-Öle-Fette-Wachse (1968) 94, p. 849

National Bureau of Standards, *Color—Universal Language and Dictionary of Names* (NBS Special Publication 440, Stock No. 003-01705-1) (1974) Washington, DC

8 Pigment Control: The Crucial Difference Between "Traditional" Versus "New" Techniques

8.1 Introduction: Energy-state Analysis

During the dispersing process, energy is added to the coating, with the intermediate result being an "energy-rich" system that may eventually attempt to transform itself into a less energy-rich system. If pigment stabilization has not occurred, then the finely dispersed pigments may build uncontrolled flocculates. Such flocculates are, structurally speaking, rather similar to agglomerates—except that the interstitial spaces in flocculates are generally filled with resin solution rather than air. Dozens of performance features are adversely affected by the presence of such flocculates. For instance—light scattering and particle size distribution can be altered in such a way that lack of reproducibility in regard to color, gloss, opacity, and other macroscopic appearance parameters may occur. As shown in Figure 8.1, the aforementioned light scattering variable can sometimes be related (in a situation-specific fashion) to particle size and/or flocculate size distribution.

How can variables such as light scattering, particle size distribution, and overall appearance be properly controlled? From the practical perspective of the coating formulator, how can one achieve "real world customer satisfaction" by manipulating the above variables? What happens inside the coating at the seemingly theoretical and esoteric level of individual pigment molecules? The answers to the above questions will be developed in a full-fledged discussion of "traditional" versus "new" pigment control techniques.

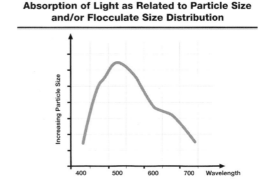

Figure 8.1 Absorption of light

8.2 Pigment Control: An Overview of both "Traditional" Versus "New" Techniques

Several different pigment control techniques have evolved over the past few decades. For the purposes of the particular discussion at hand, such techniques are classified into two main categories—*traditional* and *new*. Traditional pigment control techniques will be covered in a relatively brief fashion; however, the bulk of the following discussion will center upon new techniques. Where appropriate, the aforementioned categories will be sub-divided into key conceptual areas such as wetting, dispersing, waterborne/aqueous, solvent-based, etc.

8.2.1 Traditional Techniques: Wetting Additives

Pigment agglomerates must somehow be wetted by the binder solution. This wetting process is influenced by a wide variety of factors ranging from the geometry of the interstitial spaces in the interior of the agglomerates to the polarities of not only the pigment surfaces, but also of the binder solution itself. Especially important is the *interfacial tension* between the pigment surface and the binder solution.

Wetting additives, in the traditional sense of the word, can be defined as substances which are designed to reduce interfacial tension *within the special context of the pigment's environment*; therefore, they accelerate the wetting process through both *chemical* and *physical* mechanisms. The crucial point of interest is that mere interfacial tension reduction itself may *not* be sufficient, because performance improvement only comes about when a *balance* of pigment-related properties (optical, chemical, and application-specific) is achieved. It comes as no surprise then that some of the world's best interfacial tension reducers are very poor pigment wetters.

Traditional wetting additives often display a "surfactant-tenside" type of structure. Characteristic molecular structures may simultaneously contain polar, hydrophilic components along with non-polar, hydrophobic entities. Because of their combination-type structures, such chemical compounds orient themselves toward the pigment/resin solution interface. From a chemical viewpoint, wetting additives can be either ionic or non-ionic according to exactly how the polar segment is incorporated into the overall molecule. For illustrative purposes, the general schematic structure of a simplified wetting additive is shown, along with a randomly selected example molecular structure (Fig. 8.2).

8.2.2 Traditional Techniques: Dispersing Additives

Traditional dispersing additives adsorb upon the pigment surface, maintain proper separation between pigment particles (through utilization of steric hindrance *or*

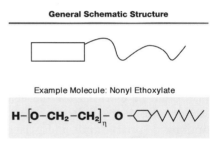

Figure 8.2 Schematic and molecular structure

electrostatic charge repulsion—but usually not through simultaneous utilization of both techniques), and therefore reduce the tendency to uncontrolled flocculate formation. The choice of optimal dispersing methodology is generally contingent upon whether the system is waterborne or solvent-based.

8.2.2.1 Aqueous Systems

Given the increasingly wide array of coating systems containing water as the carrier liquid, it is understandably impossible to characterize one absolute, "cure-all" dispersion technique for virtually all applications. Nevertheless, the best mechanism is generally a *two-pronged approach in which a traditional wetting agent is combined with an electrostatic-repulsion-providing dispersing agent (such an approach is especially appropriate for emulsion-based systems)*. The major types of chemical structures involved—as outlined in the two classical examples shown in Figure 8.3—are often polyelectrolytes, higher molecular-weight molecules in which the side-chains carry a multitude of charges.

Well-known examples, such as sodium salts of polyacrylic acids, have found widespread application by adsorbing upon the pigment surface and consequently transferring their electrical charge to the pigment. Counterions concentrate themselves in the vicinity of the pigment surface in the liquid phase, thus forming the classical electrical "double-layer". All pigment particles, though, must receive the same

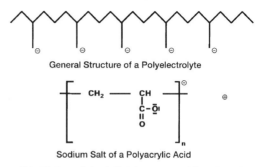

Figure 8.3 Representative structures: Polyelectrolyte and polyacrylic

electrical charge. Whether or not the charge itself is positive or negative is not always the prime point as long as all particles still receive the same charge (please note that there can obviously be exceptions to this rule). In ideal situations, flocculation is drastically reduced; however, a "Dispersant Demand Curve" (Fig. 8.4), along with extremely precise control of variables such as slurry/letdown equilibrium (Fig. 8.5), may often be essential for the achievement of optimal control. (The figures in this section also help illustrate what may possibly happen when, for instance, non-optimized slurry and letdown contain inappropriate quantities of emulsifier, dispersant, and/or other ingredients. Upon mixing, the resultant system naturally attempts to reach equilibrium. This "equilibration attempt", however, may set off a chain-reaction of physicochemical events that can, in turn, occasionally cause flocculation or coagulation of the emulsion—all of which can be avoided by the tailored usage of a proper wetting agent.)

It comes as no surprise to note that one's relatively simple theories often become more complex when one is confronted with the realities of the day-to-day world. As discussed in latter sections of this text, newly developed pigment control techniques have been designed to cope with some of the more complex situations faced by the formulator of environmentally friendly coatings.

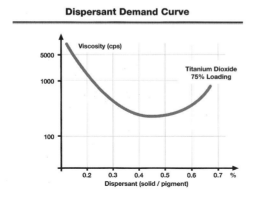

Figure 8.4 Empirical determination of dispersant demand

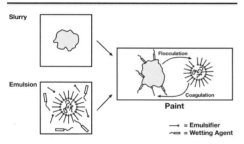
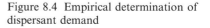
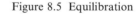

Figure 8.5 Equilibration

8.2.2.2 Solvent-Based Systems

Traditional pigment control mechanisms in solvent-based systems are generally completely different from those found in aqueous systems. First of all, effective dispersing additives for solvent-based systems distinguish themselves through the presence of the following two unique structural elements:

- "Pigment-affinic" or pigment-anchoring groups which are responsible for durable adsorption upon the pigment surface.
- Resin-compatible components (generally hydrocarbon chains) which, after adsorption of the additive upon the pigment surface, protrude into the resin solution.

The aforementioned protruding hydrocarbon chains function in the classical steric hindrance fashion; direct contact between and among the individual pigment particles is hindered so that uncontrolled flocculation can be effectively prevented.

Through the specific structures of the affinic groups (generally polar) and the resin compatible chains (generally non-polar), such traditional dispersing additives demonstrate clearly defined surfactant and wetting properties. In contrast to pigment control agents for aqueous systems, the same molecule can simultaneously do "double-duty", so to speak, as both the wetting *and* the dispersing additive.

8.3 Discussion of New Pigment Control Techniques: Wetting and Dispersing

As discussed previously, a very strong and durable adsorption of the additive upon the pigment surface is of utmost importance. What happens though when one considers a wide variety of pigment surface features, including pigment polarity? How is the adsorption affected?

In regard to inorganic pigment molecules, the surface is, of course, generally composed of relatively polar entities. So this means that the adsorption process is comparatively easy. In contrast, organic pigments (especially some of the more modern ones) display entirely different molecular structures and, concomitantly, a series of altered and more complex adsorption parameters. In practice, many of the traditional wetting and dispersing additives may not sufficiently deflocculate and stabilize the system. This problem is further compounded in environmentally friendly coatings because of the possibly higher surface tensions of resin and carrier liquid components. Consequently, new techniques which involve *three* factors—steric hindrance, electrostatic repulsion, and advanced pigment-mobility equilibration (including, where appropriate, interfacial tension modification)—are more essential than ever. In addition, as shown below, there are certain structural and/or performance characteristics which clearly distinguish the new families of pigment control products from the traditional additives discussed thus far:

- Significantly higher molecular weight of the additive itself, to the point where the additive demonstrates an almost resin-like character
- An extremely high number of pigment adhesion groups
- The ability to impart not only tailored steric hindrance, but also electrostatic charge repulsion stabilization

Products functioning on the basis of the above characteristics are especially useful in conjunction with the more complex organic pigments. The steric hindrance provided by the multiple protruding polymer chains of the additive is optimized when such chains are most compatible with the resin solution. Because compatibility is so crucial, then several different families of products have been designed so that polymer chain compatibility can, in effect, be precisely tailored to the system at hand. The germane performance features of one product family will now be summarized in Section 8.4.

8.4 Tailored Application Areas According to Molecular Weight and Polarity

As shown in Figure 8.6, optimized performance in a wide variety of application areas can be achieved by varying the characteristics of the polymeric additive.

When utilized in conjunction with *organic* pigments, the recommended usage levels of special polymeric products are significantly elevated as compared to conventional additive usage levels. In contrast, the usage levels for inorganic pigments generally match the levels to be expected when conventional additives are employed. Elevated levels (in the former case) are absolutely necessary because of the correspondingly larger surface area of organic pigments. At this point, it is important to emphasize that the relatively high usage levels of the new additive families do *not* adversely affect weather resistance and/or other coating properties.

How can one employ evaluative techniques to choose the best polymeric additive from the wide variety of products available? Obviously, one technique would be to consider molecular weight and polarity variables, but such variables alone are rather insufficient. They are vitally important, but they do not necessarily represent "the real world". Much broader factors should be incorporated into the action plan. Ideally, one should include evaluations of the following: (1) *crucial end-use optical properties of the final coating itself* (gloss, hue, color strength, and transparency), (2) *mechanical properties* (hardness and flexibility), and (3) *properties from an application point of view* (handling characteristics, sagging resistance, levelling etc.). For instance, in combination with the data in Figure 8.6, one could perhaps best optimize certain optical properties by analyzing appropriate information about color index and particle-diameter/light-scattering relationships. In this fashion, color and hue features could be properly fine-tuned. Naturally, a complete discussion of all the above variables is far

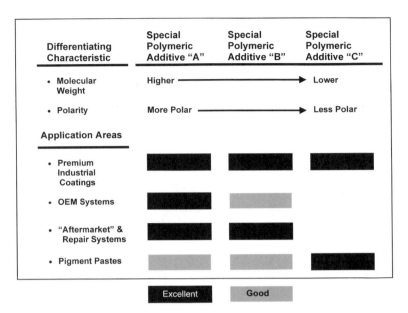

Figure 8.6 Optimized performance as related to differentiating characteristics

beyond the scope of this publication; however, Table 8.1 (in which selected excerpts from a special Red Color Index Comparison are shown) merely displays one of the many diagnostic pigment control tools utilized by formulators of certain advanced coatings. Some systems require that one meticulously correlate and match, on a closely controlled statistical basis, scores of pigment–resin–additive–solvent variables to the exact conditions of the particular formulation at hand. Additional diagnostic tools and correlation measures are shown in Figures 8.7 and 8.8. (Given the nature of this overview publication, the aforementioned tables and figures are designed for illustrative reference purposes only; therefore, with the exception of the ensuing discussion below, additional theoretical background information will not be discussed. Please note that flocculation parameters may greatly shift the critical values displayed in Figure 8.7 and elsewhere.) By examining various pigments, certain rare exceptions to some of the principles discussed thus far can be identified. For example, one occasionally encounters selected pigment types which do not always uniformly (1) deflocculate or (2) acquire positive charges when used in conjunction with certain polymeric additives. (Consequently, pigment interaction and color control become even more complex; some titanium dioxide types fall into this category.) Given the possibility of this behavior, it is important to at least mention that several special flocculation state factors must *also* be considered when addressing such performance criteria as color control. As shown in Figures 8.7 and 8.8, system-dependent factors worthy of *initial consideration* include (1) color strength and hiding power in relation to flocculation state, and (2) interactive color effects based on whether the pigments absorb or scatter light.

Table 8.1

Color Index (CI) pigment red	CI constitution no.	CI chemical class	Hue	Fastness								
				Organic solvents							NC solvents	Xylene
				Aliphatic petroleum	Cellosolve	Esters	Ethanol	Ketones				
11	12430	M. azo (pigment)	Bright bluish red (tint reddish violet)	2	2	2	1–2	1–2		1		1
12	12385	M. azo (pigment)	Bluish red (tint reddish violet)	2–3	2	2	2–3	1		1		1–2
13	12395	M. azo (pigment)	Bright bluish red	Insoluble	—	—	Excellent	—		Poor		Excellent
14	12380	M. azo (pigment)	Bluish red	Excellent	—	—	Poor	—		Poor		Poor

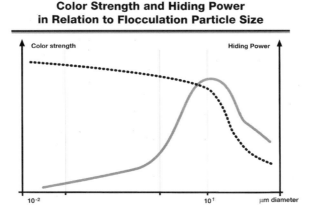

**Color Strength and Hiding Power
in Relation to Flocculation Particle Size**

Figure 8.7 Crucial variables

Interactive Color Effects

	Absorption	Scattering	
Carbon blacks	O	°	Nonselective
Whites	°	O	Nonselective
Inorganic pigments	O	O	Selective
Organic pigments	O	O	Selective

Figure 8.8 Pigment parameters

Although the above *preliminary* variables and considerations must certainly be kept in mind, the question of exactly why certain pigments do not always uniformly deflocculate or acquire positive charges still remains. What exactly might be occurring in these exceptional situations? *The key to answering this question often lies in the following concept—competition for "pigment-sites" by the solvent and/or resin.*

When an additive molecule adsorbs upon the pigment surface, the additive competes with both the solvent and the resin—all of which are vying for sites on the pigment surface. The strongest competition comes from the resin, because of its polymeric nature. On the other hand, solvent molecules can be driven away from the pigment surface quite easily. In the few situations where certain titanium dioxide pigments do not allow their surface charges to be altered, the aforementioned "competition effect" is strong enough to prevent the additive from achieving its optimal performance. How can this problem be overcome?

One method to successfully avoid the "competition effect" is to change the grinding method so that the first phase of grinding occurs *without the resin*. To be specific, a *pre-mix* technique is employed in which only the titanium dioxide, the additive, and the solvent are combined together in a dissolver or similar apparatus. Afterwards, in a separate second step, the resin is added and then grinding resumes as normal. The actual formulation itself is not altered; merely the processing stages, along with the order of component addition, are changed.

8.5 Special Factors Influencing the Rate and Quality of Pigment Wetting and Dispersing

Given the ever-increasing pressure upon both cost and performance factors in environmentally friendly coatings, it is essential to consider pigment wetting and dispersing within the overall business and technical context. For instance, whenever the rate of wetting is increased, then capital utilization and production rates are better optimized. Likewise, in an equally unsurprising fashion, even the most subtle improvements in dispersion quality can positively affect the bottom line as well as customer satisfaction. Sometimes an extra "performance edge", so to speak, is necessary to deliver the utmost in economic and technical results; accordingly, this is an opportune time to at least introduce a series of special factors influencing the rate and quality of pigment wetting and dispersing.

8.5.1 Wetting Dynamics and Color Control Considerations

In regard to most single pigment grinds, it is relatively simple to decide whether (and how) the pigments are to be transformed into respective states of either deflocculation or controlled flocculation. Unfortunately, the situation with even relatively simple pigment mixtures (either all organic or all inorganic) poses a significantly increased degree of complexity. Consider the case of color pigments which may behave differently according to (1) *the evaporation features of the overall system,* and (2) *the flocculation state.* Now suppose the situation is compounded further by the combination of *inorganic* and *organic* pigments. What can happen to color control and wetting dynamics? Color striation, flood/float, and lack of reproducibility from batch to batch may obviously occur.

Solvents* generally display lower surface tensions than resins, so this means that the surface tension of the coating system, as a whole, increases as evaporation occurs.

---

*Part of the commentary above is obviously predicated on the existence of a solvent-based system (such as a high solids coating); however, the classical Washburn equation can also be used to describe pigment wetting behavior in many waterborne systems. It is important to note, though, that the equation *must be used in conjunction with flocculation state information in order to provide a more complete depiction of pigment control.* The evaporative characteristics and surface tension dynamics of certain water-based systems do not necessarily correlate with those of solvent-based counterparts.

In practice, such evaporation almost never occurs with complete uniformity on the film surface; therefore, differences in surface tension result. Paint material is transported away from the centers of low surface tension.

In combination with the introductory variables shown previously in one version of the Washburn equation, *density* also exerts an important influence. Resins are denser than solvents, so the corresponding density in the evaporation centers must increase. When one considers the additional fact that one very crucial side-effect of solvent evaporation is decreased temperature (which, in turn, may cause a further increase in density)—it becomes clear that a very dynamic, interconnected series of events is occurring. The paint material (including pigments) which is more dense sinks in the liquid and is correspondingly replaced by more solvent-containing material from below. As a result, *vortexes, which adversely affect pigment mobility, begin to form.* As mentioned previously—color striation, flood/float, and lack of reproducibility may occur from batch to batch. In order to control pigment mobility, the threefold approach of steric hindrance, electrostatic repulsion, *and advanced pigment-mobility equilibration* (including interfacial tension modification as appropriate) must be employed. In certain extreme situations, it may also be necessary to use a polysiloxane additive in conjunction with one of the new wetting and dispersing products; obviously a complete discussion of surface and interfacial tension effects is beyond the scope of this particular publication.

8.5.2 Special Comments about Flocculation State as Related to Color Control

Are there any "rules of thumb" which are easy to apply in regard to judging the quality of dispersion? Yes, as shown in Table 8.2, the color location of the pigment generally shifts in a predictable manner according to dispersion state.

How can one explain the above observations? In terms of background information, the color of a pigment obviously results from the partial reflection of the incident white light. For many inorganic pigments, this can primarily be ascribed to *diffusion*; whereas with many organic pigments, the decisive factor is the absorption of certain wavelengths. From an optical point of view, smaller particles (and/or "particle entities" such as controlled flocculates) are often more effective than larger ones, so certain mathematical relationships can be discerned in regard to color control. For

Table 8.2   Color Shift Chart

Color	Flocculated	Deflocculated
Yellow	Reddish	Greenish
Red	Bluish	Yellowish
Blue	Greenish	Reddish
Green	Yellowish	Bluish

instance, as the particle diameter in a given system goes below 1 μ, then the color location of the pigment generally shifts counterclockwise around the achromatic point of a color ellipse. This means that the wavelength of the reflected light becomes smaller. In summary, when one is working with particular pigment varieties which change their color and color location during dispersion, there can be *only one state of optimal dispersion: complete deflocculation.*

Guide to Further Reading

Andrews, C.L., *Optics of the Electromagnetic Spectrum* (1960) Prentice-Hall, Inc., New York
Edisbury, J.R., *Practical Hints on Absorption Spectrophotometry* (1967) Plenum Press, New York
Johnston, R.M., Saltzman, M., *Industrial Color Technology* (1971) American Chemical Society, Washington, D.C.
Kuehni, L., *Computer Colorant Formulation* (1975) Lexington Books, Farnborough, England
Stearns, E.I., *Practice of Absorption Spectrophotometry* (1969) John Wiley and Sons, Inc., New York
Wright, W.D., *Measurement of Colour,* Third Edition (1964), Van Nostrand Co., Inc., Princeton, NJ
Wyszecki, G., Stiles, W. S., *Color Science: Concepts and Methods, Quantitative Data and Formulas* (1967) John Wiley and Sons, Inc., New York

9 The Interdependence of Rheological Variables and Pigment Particle Parameters

9.1 Introduction: Newly Patented Developments

Modern advances in the control of rheological variables and pigment particle parameters often allow nearly exponential performance improvement. Newly developed rheology agents can often display *2 to 20 times the efficiency of conventional products;* furthermore, high-molecular-weight wetting and dispersing additives may sometimes require—in comparison to their lower molecular-weight analogs—up to 60–70% less pigment for the achievement of equivalent color development.

Although certainly no panacea is offered in every situation, how is such dramatic improvement possible when it does occur? What are the structure–performance relationships that are responsible? In answer to these questions, an analysis of rheological variables within the context of pigment particle parameters is presented. Newly patented developments are discussed within the following theoretical and practical framework:

- Overview of structure–performance relationships
- The dual variables of rheology control and interpigment forces
- Conceptual background: The defining framework of rheology
- The keys to performance improvement with molecular networks

9.2 Overview of Structure–Performance Relationships

Two of the more complex phenomena in the study of liquid film technology are rheological variables and pigment particle parameters. Thus far, a comprehensive study of the complex interrelationships between these two phenomena has not yet been undertaken in the coatings field. Accordingly, the objective of this publication is to take at least one step forward in this arena—*to fill part of the gap*—namely, *to identify and study some of the basic building blocks and interrelationships.*

In particular, the objective will be, first of all, to introduce newly patented chemistry and explain how certain rheology agents can often display 2 to 20 times the efficiency of conventional analogs. Secondly, in the course of explaining how the above rheology agents can so dramatically improve performance, the concomitant technological developments in the high molecular-weight wetting and dispersing additives field will be touched upon. How and when can one harness, for instance, *"molecular networks" (provided by specially modified urea-based rheology agents)* to dramatically improve performance? How and when can one produce controllable cohesive interactions which routinely culminate in, for example, 300 to 400 + % improvement in anti-sag and anti-settling properties? Answers to the above questions will be provided in the remainder of this chapter.

9.2.1 Basic Chemical Structures and Advanced Molecular Networks

With the objective of laying the groundwork for the physicochemical relationships responsible for performance improvement, four key introductory figures must be examined. Accordingly, the absolute first step in an integrated study of rheological variables and interpigment phenomena is a cursory overview of the aforementioned figures: Figure 9.1 displays the basic chemical structure of a new rheology-modifying compound, and Figure 9.2 graphically overviews the cohesive molecular network structures formed by this compound. Furthermore, in accentuation of the performance improvement imparted by this new chemistry, Figure 9.3 depicts a practical example of at least 300 to 400% performance improvement. *How is such performance enhancement possible? What is different about this new chemistry?* The crucial difference is as follows: Modified urea products are not relegated to influencing rheology by means of traditional mechanisms such as controlled flocculation. (Please see Figure 9.4, where a traditional rheology control mechanism is depicted.) Modified ureas employ advanced molecular networks which utilize strong and highly controllable cohesive interactions.

9.2.2 300 to 400 + % Performance Improvement

Compared to conventional products, the modified urea moiety in Figure 9.3 provides significantly enhanced performance, often *at only one-quarter the usage level* of conventional analogs; accordingly, when one considers the one-quarter usage level factor, an equivalent of at least 300 to 400% performance improvement occurs. Such

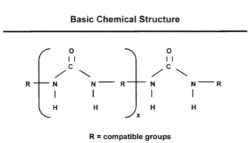

Basic Chemical Structure

R = compatible groups

Figure 9.1 Structure of rheology-modifying compound

New Technology
Advanced Molecular Networks

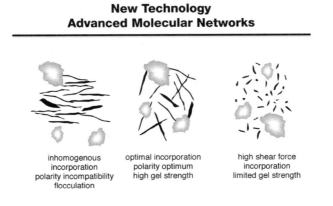

| inhomogenous incorporation polarity incompatibility flocculation | optimal incorporation polarity optimum high gel strength | high shear force incorporation limited gel strength |

Figure 9.2 Graphic depiction of cohesive networks

300 - 400+% Performance Improvement
The Enhanced Anti-Sag Properties of Molecular Networks

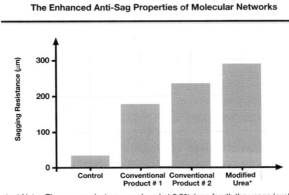

*Important Note: The urea product was employed at 0.3% (one-fourth the usage level of competitive or conventional products); reproducible performance enhancement results of similar magnitude were obtained in such diverse systems as acrylics, urethanes, alkyds, expoxies, etc.

Figure 9.3 Magnitude of performance as related to differentiating characteristics

Traditional Performance Control Mechanism
Pigment Adhesion Groups

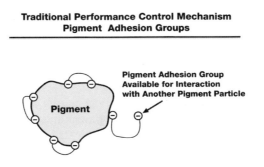

Figure 9.4 Comparative mechanism

magnitudes of performance enhancement (tripling or quadrupling) are relatively common, and in selected instances, rheology improvements of up to 20 times the efficiency of conventional products can be achieved. (Please note that, for purposes of brevity, an entire compendium of graphs is not shown in this chapter. Nevertheless, dozens of additional graphs, many depicting substantially more dramatic performance improvement than that shown in Figure 9.3, are available.)

9.3 The Dual Variables of Rheology Control and Interpigment Forces

9.3.1 Introduction to Molecular Networks

Certainly a multivolume treatise would be required to properly delineate all of the true complexities involved in rheology and pigment wetting phenomena, so the focal point of this overview publication will center primarily on how one can achieve nearly exponential improvement of rheological properties. Obviously, as an important sideline, the concomitant improvement of wetting and dispersing properties will also be discussed as appropriate; however, the prime focal point will be *molecular networks*.

With the above objective in mind, Section 9.3.2 of this chapter will introduce *the concept of pigment adhesion groups and their critical influence upon rheology properties*. The sometimes underestimated differences between purely deflocculating additives and their controlled flocculating counterparts will be briefly described, with the intent of properly setting the stage for a definitional framework of rheology (discussed in Section 9.4). Without doubt, the special molecular networks utilizing modified urea moieties provide the most optimal performance, so important *supplemental data— including constant shear stress (CSS) and constant shear rate (CSR) viscosity profiles* will be presented in Section 9.5.

In addition, the further performance advantages of molecular network products (liquid form, ability to be post-added, etc.) will also be overviewed as appropriate.

9.3.2 The Pigment Adhesion Group Concept—Deflocculation Versus Flocculation

Of course, it is no secret that, when purely deflocculating additives are employed, the highest gloss and the most optimal color development and reproducibility (no flooding and floating) can be achieved. In terms of molecular structure, deflocculating additives are distinguished from controlled flocculating additives by an "overabundance", so to speak, of pigment adhesion groups. Rather than containing merely the isolated pockets of pigment adhesion groups displayed in Figure 9.4, deflocculating additives may contain 12 or more high molecular-weight adhesion groups.* As a result, the rheological behavior (reduction of thixotropy and flow barriers, resulting in almost Newtonian flow) that is achieved with purely deflocculating additives promotes excellent levelling. In addition, the commonly described vortex currents, along with flooding and floating phenomena, can sometimes be alleviated.

In contrast, the best anti-settling (and anti-sagging) behavior can be achieved when the judicious usage of controlled flocculating additives leads to the formation of special three-dimensional structures *based on pigment adhesion groups*. As a result, thixotropic flow behavior occurs. In a resting state, the viscosity is rather high; however, upon application of shear forces, the *conventional* three-dimensional structures (controlled pigment flocculates) break apart and therefore provide lower viscosity. Afterwards, upon removal of the shear forces, the controlled flocculates rebuild.

In comparison to conventional products, modified ureas demonstrate considerable advantages. For instance, *advanced* molecular networks, in contrast to the conventional structures outlined above, *can even function in systems containing no pigments!* Of course, the advanced products interact in a very beneficial fashion with pigments, at least when such pigments are present—but advanced urea-based derivatives can even function independently of both pigments and pigment adhesion groups. Additional details of the functioning mechanism exhibited by modified ureas will be provided in Sections 9.4 and 9.5 of this chapter.

Further differences exist between conventional products and their more advanced urea-modified counterparts. For example, when utilizing conventional rheology agents, the prevention of sagging and settling requires not only the *absence* of *direct pigment/pigment contact*, but also requires the *presence* of specialized pigment adhesion groups within the molecular structure of the wetting and dispersing additive. With controlled flocculating additives, one unique feature of the pigment adhesion groups is that they are capable of interacting *not only with the pigments, but also with other pigment adhesion groups within the system*. In essence, a "bridging" effect occurs, in which pigments are bound together by pigment adhesion groups in a controlled three-dimensional network bordered by pigments. Extraneous variables, such as the classical pigment/vehicle flocculation effects, are effectively removed. This is the true key to the traditional methodologies which control both settling and sagging behavior. In contrast, though, advanced molecular networks (such as the example shown previously in Figure 9.2) are *not limited* by the bridging requirement of pigment adhesion groups. *As a result, the three "R" groups (coatings compatible linkages) displayed in Fig. 9.1 can freely interact, thus resulting in a controlled formation of fine*

---

*High concentrations of pigment adhesion groups in *purely deflocculating, high molecular-weight products* can be responsible for very dramatic performance improvements, especially when 12 to 14 pigment adhesion groups are employed. In particular, up to 60–70% less pigment may be required in selected instances for the achievement of equivalent color development. For optimal performance, additives with multiple adhesion groups should, as appropriate, be used in combination with rheological additives of the type described in this chapter.

needle-like structures demonstrating strong molecular forces—primarily adhesive in nature. The end result is that, as mentioned above, the urea-modified product can even function entirely independent of the presence of pigment!

In further accentuation of the advantages offered by the urea functionality, the presence of a pigment is easily accommodated into the advanced molecular network. For instance, anti-settling behavior, along with improved sag performance on vertical substrates, can be dramatically improved. Some manufacturers even use urea-modified chemistry to help stabilize the difficult-to-handle, "special-effect pigments" utilized in the automotive industry. The urea-modified moiety is recommended for producing special-effect pigment slurries which are storage-stable, non-settling and pumpable. The new chemistry represented by the urea functionality lends itself to the production of a liquid thixotrope, which can be employed as a stand-alone rheology additive. This additive not only improves the anti-settling behavior of pigments in the slurry and in the ready-made coating (circulation pipe stability is also enhanced), but also improves orientation of pigments during application (better "flip/flop"; less "cloudiness" and haze).

9.4 Conceptual Background: Viscosity and Rheology

9.4.1 An Essential Definition

One of the most important prerequisites to an understanding of advanced molecular networks is, quite simply, an understanding of the defining framework of rheology. In and of itself, rheology is not a very complex concept at all; however, the semantics and

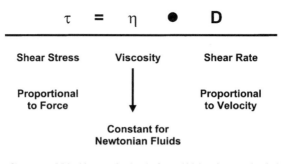

$$\tau \quad = \quad \eta \quad \bullet \quad D$$

Shear Stress	Viscosity	Shear Rate
Proportional to Force		Proportional to Velocity

Constant for
Newtonian Fluids

Note: Shear stress "τ" (tau) is proportional to the force which is acting upon the liquid. Shear rate "D" is proportional to the flow velocity of the liquid. Viscosity "η" is only a constant for the ideal case of a Newtonian fluid. In many practical cases, it is not constant, but depends upon the shear rate or on the shear time.

Figure 9.5 Viscosity definition

terminology utilized in discussions of rheological performance may be situation-specific and may even vary greatly in usage from one branch of industry to another. *Consequently, a very cursory overview of basic concepts is included, primarily so that a uniform understanding of the information presented in Section 9.5 ("The Keys to Performance Improvement with Molecular Networks") can be developed.* To begin with, viscosity is commonly defined as "the relationship of shear stress to shear rate". Of course, the viscosity relationship itself is oftentimes expressed as shown in Figure 9.5.

9.4.2 Equations and Relationships

Rheological principles can be described within the conventional context of a "two-plate model" in which, as shown in Figure 9.6, two plates are placed at a specified distance "*h*" from one another. Generally, the lower plate is fixed in place, however the upper plate is allowed to move with a speed (or velocity) of "*v*".

The left half of Figure 9.6 illustrates a theoretical case of elastic behavior, regardless of whether one is utilizing pigment adhesion groups or molecular networks to control rheology. Suppose that a particular "solid" body is placed between the two plates. Upon movement of the top plate, this solid is deformed; nevertheless, when the force (*F*) is removed, the solid regains its original shape. Such behavior can be described as "elastic".

On the other hand, what happens when a typical fluid takes the place of the solid body? As shown in the right half of Figure 9.6, the liquid begins flowing as force is exerted. Two widely accepted values, which help describe the physical process that occurs, are shown in Figure 9.7. (The reader will note, of course, that these same two values can be used—in the fashion described previously—to define "viscosity" itself.)

Viscosity changes over time and with shear rate, so many graphical depictions can be created to describe the behavior of coating systems. (Please note that some branches of the industry may even employ various definitions and nomenclature variants—sometimes conflicting—of certain viscosity-related terms.) The presence of pigments

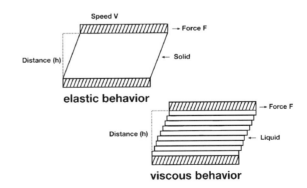

Figure 9.6 "Two-plate" model

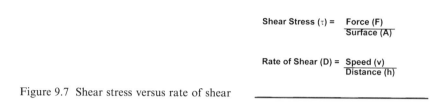

Figure 9.7 Shear stress versus rate of shear

with different surface characteristics and features creates many complex flow behavior charts. In order to properly set the stage for both the CSS/CSR data and the concluding commentary in Section 9.5, explanatory diagrams of an overview nature—depicting classical pseudoplasticity, dilatancy, thixotropy, and Newtonian behavior—are included for reference purposes as Figures 9.8 and 9.9.

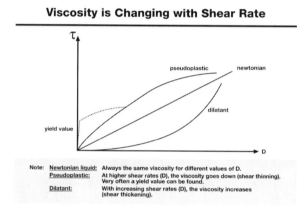

Figure 9.8 Special viscosity relationships

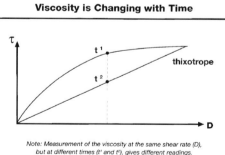

Figure 9.9 Viscosity versus time

9.5 The Keys to Performance with Molecular Networks

9.5.1 Viscosity Profiles

Given the defining framework of rheology discussed in the previous section, additional data about the rheological performance of urea-modified products can now be presented and placed in its proper context. As shown in Figures 9.10 to 9.13, the resultant viscosity profiles (including both constant shear rate and constant shear stress profiles performed at 20 and 80°C, respectively) display superior performance with the advanced urea derivative.

9.5.2 Shear Rate Versus Effectiveness

Especially under rather strenuous conditions, such as post-addition scenarios and in situations where high temperatures are encountered, the urea product proves even more advantageous than ever. As shown in the final summary figure, "Viscosity

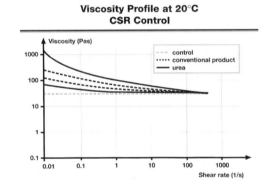

Figure 9.10 Viscosity profile #1

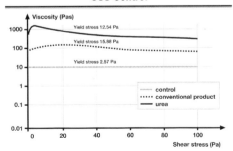

Figure 9.11 Viscosity profile #2

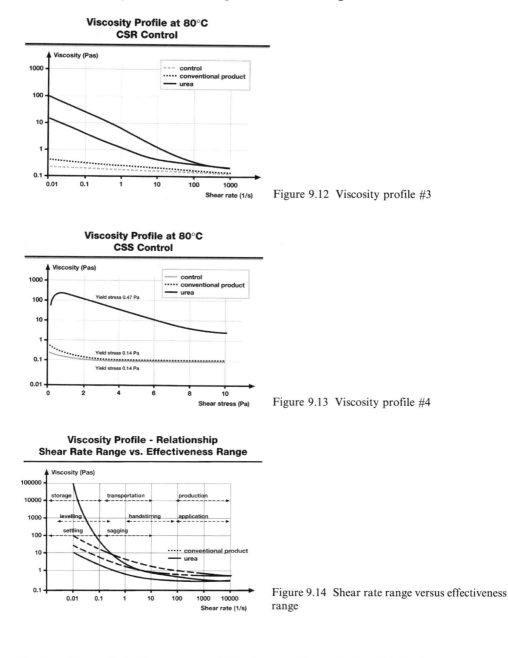

Figure 9.12 Viscosity profile #3

Figure 9.13 Viscosity profile #4

Figure 9.14 Shear rate range versus effectiveness range

Profile—Shear Rate Range versus Effectiveness Range" (Fig. 9.14), the pronounced pseudoplastic properties (shear thinning) and true thixotropy demonstrated by the advanced product are superior. Viscosity increase is quite dramatic at lower shear rates, but at elevated shear rates (where performance often matters the most) little or no viscosity increase is observed.

In conclusion, the urea-modified product provides optimal performance during storage, transportation, production, and application. Both rheological variables and pigment particle parameters are enhanced. Furthermore, the product is liquid and can be easily post-added.

Guide to Further Reading

Billmeyer, F.W., Saltzman, M., *Principles of Color Technologies* (1966) John Wiley and Sons, Inc., New York

Dallavalle, J.M., *Micromeretics—The Technology of Fine Particles* (1948) Pitman Publishing Co., New York

Degussa Corporate Communication, *Degussa-Produkte für Lacke und Farben, Schriftenreihe Pigmente* (1970) Nr. 10, 5., überarb. Auflage

EPS 0115585, USP 4696761, USP 4314924

Kerker, M., *Scattering of Light and Other Electromagnetic Radiation* (1969) Academic Press, Inc., New York

Lubs, H.A. (Ed.) *Chemistry of Synthetic Dyes and Pigments* (ACS Monograph) (1965) Hafner Publishing Co., New York

van Olphen, H., *Introduction to Clay Colloid Chemistry* (1977) John Wiley and Sons, Inc., New York

Orr, C., Jr., Dallavalle, J.M., *Fine Particle Measurement* (1959) The Macmillan Co., New York

Orr, E.W., Proceedings of the Association of Formulation Chemists, Las Vegas, September 2–4 (1997)

Patton, T.C., *Paint Flow and Pigment Dispersion* (1964) Interscience Publishers, Inc., New York

Pratt, L.S., *Chemistry and Physics of Organic Pigments* (1947) John Wiley and Sons, Inc., New York

Wapler, D., Über die Messung des Fliessverhaltens von Anstrichmitteln, *Deutsche Farben-Zeitschrift* (Jan/Feb 1958)

Zorll, U., Lichtmikroskpische Oberflächenuntersuchung von Lackschichten mittels einer Abdruck-methode, *Deutsche Farben Zeitschrift* (1961) 8

Zorll, U., Untersuchungen des viskoelastischen Verhaltens von Anstrichfilmen, *Farbe + Lack* (1965) 1

10 How to Produce Enhanced Pigment Concentrates

10.1 Introduction

Many recent technological advances have occurred in the field of rheology and dispersion control. Important new and patented developments have centered upon the realm of performance enhancers for the purposes of deflocculation and controlled flocculation.

Such performance enhancers function in an integrated fashion and can affect a wide variety of factors ranging from microscopic steric hindrance and electrostatic charge repulsion to macroscopic gloss, color development, and overall film performance.

10.2 Pigment Pastes: Theory and Application

As shown in Sections 10.2.1 to 10.2.2, pigment pastes and concentrates are utilized for two primary purposes:

10.2.1 For Tinting and Shading

An already "complete" coating system (with the pigment grinding and dilution phases consummated) can be "fine-tuned", in terms of color development, with pigment pastes.

10.2.2 For "Full System" Production

A coating system can be formed from a mixture of pastes which will be subsequently reduced or combined with the desired resin. Naturally, this scenario requires higher quantities of pigment pastes than in mere "tinting" or "shading" applications. Because the pastes exert such vast influence upon the final paint quality, then the corresponding quality demands placed upon the pastes themselves are understandably quite high.

Although the utilization of pastes as tinting and shading media has long been accepted practice, the usage of pastes in the production of full coating systems has gained increasing importance only during recent years. There are two possible formulation variations in regard to the latter usage scenario. The *first* possibility is the development of a series of pigment pastes with usage in only one paint system foreseen. In such a case, the actual paint system resin would also be employed as the grinding resin for the paste. Obviously, such pastes can be optimally tailored for usage with the particular paint system at hand. The *second* possibility consists of designing a series of "universal" pastes which are compatible with as many resins as possible. Of course, "universal" pastes cannot yield optimal performance for virtually every resin; compromises are inevitable.

This chapter will describe, with the inclusion of illustrative formulations and application data, how to enhance the performance of those particular paste systems designed for the broadest and most universal applicability. The performance enhancements to be described, however, also exhibit general applicability in the more "restricted" paste systems.

Pigment pastes for full blends are generally employed in usage scenarios which require a large number of relatively small *individual* paint quantities to accommodate a wide spectrum of hues and resin systems. (Example scenarios would include many high quality industrial, OEM, auto, and related end-use applications.) Manufacture from pastes is faster than ground-up production, and consequently the coatings manufacturer can react more flexibly to customer wishes. Whether or not such a procedure is, in actuality, more cost-advantageous, must be decided by the manufacturer on a case-by-case basis.

Oftentimes pigment pastes find application in production arenas where higher degrees of automation are desired. Pigments and fillers are generally the only ingredients which are not liquids; accordingly, they cannot be optimally handled in fluid-based production systems which are otherwise completely automated. Nevertheless, if one relies upon pumpable pastes, then nothing stands in the way of fully optimized automation.

10.2.3 Ingredient Selection Parameters

Pigments: The choice of a cost-effective, high performance pigment is crucial. Since the intent is to utilize the paste as the prime medium for incorporating pigment into the final paint system, then the pigment content should obviously be as high as possible. As described in subsequent sections, high pigment content, though, can create many technical challenges in regard to rheology and dispersion control.

Grinding Resin(s): The next important component is the grinding resin. It should be selected on the basis of optimal compatibility, and should represent as small a portion of the paste as possible.

Solvent(s)*: In order to augment the solvent portion normally used for grinding, additional solvent is generally necessary to provide a workable paste. As demonstrated oftentimes in practice, the correct solvent choice can positively influence storage stability and is therefore an important consideration.

Additive(s): Specially designed additives are essential components of modern paste systems. Deflocculating additives, for instance, can be used to reduce viscosity and to achieve the highest possible pigment loading. (Tables 10.1 to 10.3 provide a preliminary summary of selected commentary on the basic ingredient categories.)

Table 10.1   Pigment Concentrates for Aqueous Systems

• Water soluble grinding resin	
• Wetting and dispersing additive	
• Water	
• Co-solvent	To avoid drying-out of the concentrate
• Pigment	
• Silicone defoamer	To prevent foam formation during pigment grind; selection of the defoamer depends on the requirements of the final paint
• Anti-settling agent	Bentonites or fumes silicas to avoid the settling of inorganic pigments

Table 10.2   Glycol/Universal Concentrates

• Pigment	
• Wetting and dispersing additive	
• Propylene glycol	
• Water	
• Defoamer	Selection of the defoamer depends on the requirements of the final paint

Table 10.3   Solvent-Based Pigment Concentrates for Industrial and Architectural Coatings

• Grinding resin	The grinding resin must display excellent compatibility
• Wetting and dispersing additive	The additive must display excellent compatibility with the grinding resins and binder systems used
• Solvent	
• Pigment	
• Anti-settling agent	Bentonites or fumed silicas to avoid the settling of inorganic pigments

*Includes other carrier liquids, such as water.

In certain special cases, it is possible to produce pure solvent pastes (with no grinding resin whatsoever). The advantage is that such pastes eliminate the possibility of compatibility problems between grinding and reducing resins. However, solvent pastes can demonstrate unique problems of their own in regard to stability and incorporation. Because of the "missing" resin component, such pastes are highly prone to drying out and, upon dilution, may be subject to shock.

10.3 Performance Requirements

As shown below, many performance criteria must be simultaneously balanced in order to provide either suitable tinting/shading pastes or optimized full system pastes:

- Stability (no settling and/or syneresis)
- No seeding or flocculation
- Proper rheological behavior
- Highest possible pigment content
- Lowest possible resin content
- Easy of incorporation, and excellent resin compatibility
- No negative influence on durability
- Simple production
- No drying out (or easy resolubility if drying out occurs)
- Fluidity (must be pumpable)

To fulfill the above requirements, polymeric deflocculating additives (which function on the basis of both steric hindrance and electrostatic repulsion) can provide control of dispersion and rheology. However, before structure–performance criteria are discussed, it is important to summarize exactly *how to test* pigment concentrates themselves. Initial procedures, along with multiple formulations, for full blends and tinting shades are discussed in the next section.

10.3.1 Case Study: How to Test Pigment Concentrates in the Laboratory

Pigment concentrates with inorganic pigments
The paste binder, the solvent, and the wetting and dispersing additive are weighed out and first mixed with a spatula or a dissolver. Then, the Aerosil and the pigment are added separately and mixed with a dissolver; the glass beads are added after pigment incorporation.

Dispersing conditions:
Dispersing time: 40 min; however, for TiO_2, 30 min
Dispersing temperature: 40°C
Velocity: 18 m/s

Pigment concentrates with organic pigments
All components of the formulation, including the glass beads, are weighed out in the following order: paste binder/solvent/additive/pigment. All components are briefly premixed with a spatula, and then dispersed with a dispermat (21 m/s; 60 min for pigments such as carbon black).

Testing of the full blends used as full mixing pastes and tinting shades (for example ingredients, see Formulation "V" in Table 10.4)
1. Addition of the paste to the clear coat, resin, or varnish, so that the total formulation contains 8% of the non-TiO_2 inorganic pigment, or 4% of the organic pigment, or 20% of TiO_2.
2. Pouring the paint onto slanted glass or a similar surface. (Paint viscosity: Ford cup 4; 25 s.)
3. Measurement and classification of both gloss and transparency.
4. Mixing the white paint with tinting paste (Mix 5 min with the shaker.)
   Mixing ratio: 100 parts white paint: 7 parts tinting paste; for carbon black, 1.5 parts carbon black paste.
5. Pouring the final product onto glass, smooth plastic sheeting, paper, or metal.
6. Use of the rub-up test for classification of shade differences.

Testing of tinting shades (for example ingredients, see Formulation "A" in Table 10.5)
1. Mixing the white paint with the tinting paste in a 10:1 ratio.
2. Production of 2 samples, one of which should be mixed 2 min, the other 10 min, with the shaker.
3. Parallel drawing of the two samples onto a coated contrast chart with a 150 μm draw-down bar.
4. Use of the rub-up test for classification of shade differences.

10.4 The Possible Effects of Different Titanium Dioxides

Selection of titanium dioxide type may exert an important influence on any particular formula's flood/float properties. Overall pigment performance is influenced by three factors—interactive particle geometry, the pigment/resin interface, and the surface treatment and chemistry of the solid particle. If, however, one were to consider the case of examining several different titanium dioxides, then the most decisive factor of all would often be surface treatment alone.

Single pigment grinds have an advantage, in that, by addition of wetting and dispersing additives, the optimal dosage can be chosen comparatively easily. The additive controls the mobility of the pigment particles during the drying and/or curing process. Thus, the wetting and dispersing additive effectively prevents pigment separation; nevertheless, even in relatively simple grinds with single pigments (for instance, grinds containing routinely used titanium dioxide), laboratory tests demonstrate quite clearly that various titanium dioxide grades and/or types can

Table 10.4 Formulation "V", Starting Point Formulation for Pigment Concentrates

Full blends with polymeric wetting and dispersing additive "C" and aldehyde resin

Pigment	Titanium dioxide	Organic yellow	Iron oxide yellow	Chrome yellow	Organic red	Molybdate orange	Iron oxide red	Organic Red–Violet	Phthalo green	Phthalo blue	Carbon black
Color index (CI)	PW 6	PY 154	PY 42	PY 34	PR 170	PR 104	PR 101	PR 88	PG 7	PB 15	PBK 7
1. Typical aldehyde resin	14.0	15.6	12.5	13.8	11.7	13.3	11.5	13.3	12.0	18.0	15.3
2. Methoxypropylacetate	15.6	21.9	18.2	9.4	23.3	7.7	7.9	26.7	41.3	43.1	28.0
3. Polymeric wetting and dispersing additive "C" (45%*)	2.2	17.5	7.0	8.7	20.0	11.5	12.2	15.0	16.7	13.9	22.7
4. Aerosil 200	0.2	0	0.3	0.6	0	0.5	0.4	0	0	0	0
5. Pigment	68.0	45.0	62.0	67.5	45.0	67.0	68.0	45.0	30.0	25.0	34.0
	100.0	100.0	100.0	100.0	100.0	100.0	100.0	100.0	100.0	100.0	100.0

*Non-volatiles

Procedure: Premix ingredients 1 to 3; then add Aerosil and pigments. The pigment content is based on the selection of the particular pigment at hand. Of course, the different pigments may display variations in degree of efficiency and also in type of rheological behavior.

Table 10.5 Formulation "A", Starting Point Formulation for Pigment Concentrates

Tinting shade with special wetting and dispersing additive "E"

Pigment	Titanium dioxide	Organic yellow	Iron oxide yellow	Chrome yellow	Organic red	Molybdate orange	Iron oxide red	Organic Red–Violet	Phthalo blue	Carbon black
Color index (CI)	PW 6	PY 154	PY 42	PY 34	PR 170	PR 104	PR 101	PR 88	PB 15	PBK 7
1. Typical aldehyde resin	2.5	2.5	5.0	7.0	2.6	7.0	5.0	7.5	2.5	8.0
2. Butyl glycol	17.6	25.3	18.4	6.4	23.7	4.6	7.5	17.3	33.8	19.5
3. Propylene glycol	17.6	25.3	18.4	6.4	23.7	4.6	7.5	17.3	33.8	19.5
4. Special wetting and dispersing additive "E" (43%*)	2.0	16.9	17.7	14.5	20.0	15.1	14.5	17.9	11.9	25.0
5. Aerosil 200	0.3	0	0.5	0.7	0	0.7	0.5	0	0	0
6. Pigment	60.0	30.0	40.0	65.0	30.0	68.0	65.0	40.0	18.0	28.0
	100.0	100.0	100.0	100.0	100.0	100.0	100.0	100.0	100.0	100.0

*Non-volatiles

Procedure: Premix ingredients 1 to 5; then add pigments under agitation. The pigment content is based on the selection of the particular pigment at hand. Of course, the different pigments may display variations in degree of efficiency and also in type of rheological behavior.

exhibit significantly different flood/float behaviors. (Of course, the propensity to exhibit different behaviors is generally not observed until the various titanium dioxides are subsequently utilized in pigment mixtures.) Accordingly, a test battery of different pigment permutations and combinations is always of interest. In one representative study, several pigments (white, red, blue, and black) were therefore studied. Different white pastes, based on different titanium dioxides (please note that all the titanium dioxides were so-called "competitive offsets" for one another), were alternately mixed with red, blue, or black to create a pastel pink, a light blue, and a medium gray. The resultant tests displayed, however, *vast color development differentials* when different titanium dioxide "single grinds" were mixed with the very same color pigments. This, therefore, demonstrates that the selection of titanium dioxide types—and also the concomitant stabilization of colored pigments—can exert a decisive influence on flood/float behavior. For instance, if there is no flooding/floating with a particular titanium dioxide type in system "X", this does not necessarily imply that the same titanium dioxide type will perform in a congruent manner in system "Y". As a corollary, if a particular titanium dioxide displays copious flooding/floating in one master system, this does not necessarily mean that it will perform in the same manner with other colors of the system.

It is often desirable to disperse various colors of pigments separately with unique, high-molecular weight deflocculating additives. Various titanium dioxide types may need to be dispersed with more than one additive for optimal results. Of course, the advantages of charge control and the "premix" method must always be kept in mind.

10.5 How Pigment Concentrates Can Be Utilized to Contain Formulation Costs and to Enhance Performance

The increasing number of coatings systems with ever-higher production costs forces the industry to find methodologies for economic rationalization. Specially prepared pigment concentrates often provide the key to success. The first crucial advantage offered by pigment concentrates is that of dual application areas (simultaneous utility in both "tinting paste" and "full mixing system" scenarios); the second crucial advantage is the overall achievement of higher performance standards, regardless of application area. Of course, both advantages result in reduced costs and improved asset utilization. *The production of specialized pigment concentrates therefore offers the coatings industry an important avenue of approach for the "reengineering" of existing production processes. This often means, however, a change in thinking and also a change in production techniques.*

Modern paint production utilizing pigment concentrates rests upon the principle of producing intermediates—in this case, pastes—for stock. These pastes can then be added during the actual production process, or can easily be added to the resins or to other components of the formulation. For factories using computerized mixing systems, pigment concentrates are a necessity for producing coatings with a wide, reproducible color range. Even though paint production with pigment concentrates finds its major utility in the manufacture of colored paints, it is also widely used for producing white systems. (Of course, even specialized extender pastes can contribute to cost savings and performance enhancement, depending upon system requirements.)

Augmented information regarding the two aforementioned cost advantages is discussed below:

- *Cost and performance advantage number one: Dual application areas*
  To begin with, pigment concentrates demonstrate dual utility in the production process:
  1. *For Tinting (Tinting Pastes)* After production of the paint by grinding the pigments and by letting down, tinting pastes are used for fine-tuning the color. A maximum of 10% tinting paste is added.
  2. *For Full Mixing Systems* A paint can also be produced by using different pigment concentrates and can then be let down by the appropriate resin (or clear coat). The procedure demands, of course, rather large quantities of concentrate. Full mixing systems display two economically efficient options. First, one can formulate a range of concentrates for merely one coating system. The original paint binder is then used as the grinding resin for the pigment concentrates as well. Second, one can create a so-called "universal" range of pastes compatible with as many binders as possible and therefore suitable for the production of different paint qualities.

- *Cost and performance advantage number two: The achievement of higher performance standards*
  As discussed below, pigment concentrates not only help reduce costs, but also *improve coating system performance* when they employ high molecular-weight additives (often containing 12+ pigment adhesion groups) to meet the following stringent performance criteria:
  1. Color homogeneity and *batch-to-batch reproducibility*
  2. Stability during storage, usage, and application
  3. Ease of preparation and handling
  4. Absence of deleterious influences on coating system performance
  5. Cost-effectiveness
  6. Flexible application areas

To significantly enhance performance, it is absolutely necessary to use high molecular-weight additives in concentrates. Only a tailored pigment wetting and dispersing additive with multiple pigment adhesion groups can reduce the paste viscosity so that the pigment content can be drastically increased. Pigment loadings with an additive versus without an additive can indeed be radically different. Stabilization of the pigment grind by additives also prevents shocks during let-down and mixing of the concentrates. To avoid flood/float as much as possible, pigments should not only be deflocculated, but the mobility of the pigments should, in effect, be homogenized through "charge-equilibration".

10.6 Usage Levels, Points of Addition, and Practical Formulation Hints

10.6.1 Determining the Appropriate Usage Level: Initial Considerations

The ascertainment of the proper usage levels of wetting and dispersing additives obviously depends, to a large extent, on pigment particle size; however, additional factors must also be considered. For instance, *if the pigment concentrates are to be used as tinting shades, then the base paint must simultaneously be stabilized to achieve total system color stability.* Accordingly, the selection of a suitable wetting and dispersing additive can be integrally related to the binder system of the base paint. Generally speaking, high molecular-weight wetting and dispersing additives are nearly always preferable. Experience has shown that, without titanium dioxide stabilization, color-stable paints cannot be formulated. Right from the beginning, a tailored wetting and dispersing additive, which prevents flocculation, should be used to stabilize the TiO$_2$. (After the base paint and the concentrates are formulated, it is then nearly impossible to correct flooding and floating which may subsequently arise from unstable pigments.)

10.6.2 Usage Levels: Basic Principles

Since an additive is overtly designed to "adhere" or attach itself to the pigment surface, then the amount of required additive depends upon exactly how much pigment surface is present. Classical wetting and dispersing additives based upon low molecular-weight polymers should be utilized in quantities of 0.5 to 2.0% with inorganic pigments and in quantities of 1.0 to 5.0% with organic pigments (additive delivery form based upon the pigment weight). Based on the entire formulation weight, additive usage levels of 0.1 to 1.0% are generally employed.

In comparison, considerably higher usage levels of the high molecular-weight additives are necessary, especially in conjunction with small particle organic pigments. Such levels are required since pigment surface area with organic pigments is much greater than with inorganic pigments. High levels are no cause for concern though, since the resin-like properties of the polymeric additives do not negatively influence coating durability. (This, of course, does not preclude routine evaluation of coating properties by the formulator.) When high molecular-weight additives are employed with inorganic pigments, the recommended usage levels can sometimes be reduced (because of the smaller surface areas involved) to levels equivalent to those generally encountered with the low molecular-weight products. Typical usage levels for inorganic pigments are 1 to 30%; for organic pigments, usage levels are 30 to 90% (additive delivery form based upon pigment weight). With very fine particle pigments—for example, some carbon blacks—usage levels of 60 to 120% are necessary.

As important criteria for the evaluation of proper stabilization—gloss and transparency should be considered. In addition, flooding and floating behavior should be evaluated with the "rub-up" test. When laboratory studies are scaled up to production levels, one must be careful to assure that equivalent grinding conditions are observed. Only in this fashion can comparable grinding results be achieved.

10.6.3 Point of Addition/Order of Addition

Wetting and dispersing additives belong in the grind so that performance can be optimally developed. The exact formulation of the grind ("resin-poor" versus "resin-rich") and the addition order of the individual components can certainly influence the grinding quality. Theoretically, the best results can be achieved when—first of all— *only the pigment(s), solvent(s), and additive are mixed together ("pre-mix")*. This allows the additive to preferentially attach to the pigments without having to compete with the resin polymers. In practice, however, this method proves necessary only for "worst-case scenarios".

Stringent demands are sometimes placed upon additives to perform in "post-add" situations and also in finished batches to correct flood/float and/or flocculation problems. Not all additives are equally appropriate for such usage. As a rule, though, rather high post-add dosages are necessary.

10.6.4 Single Grinds and Co-Grinds

Whenever only one pigment must be dispersed, then all important parameters (additive quantity and grinding conditions) can be fully optimized; the end result is therefore the best possible grind quality. In practice, however, co-grinds with several pigments are generally the rule. Compromises in regard to grinding parameters are unfortunately necessary, which means that the results cannot always be compared to those of single grind systems. In any event, one should closely evaluate the respective flocculation states of all individual pigments in the more complex systems (in order to identify "difficult-to-handle" pigments). Additional corrective measures should be determined on an individual basis; for example, problematic pigments can be exchanged for other pigments, can be ground separately, or can perhaps even be added as pigment concentrates.

10.6.5 Rheological Additive Combinations

As described previously, wetting and dispersing additives can influence paint system rheology; such influence, however, is generally only a side-effect of additive usage. For optimal rheological control, wetting and dispersing additives should be combined with other rheological additives (fumed silicas, bentonites, and dehydrated castor oils). Synergistic effects can be observed in many cases. Organoclays (bentonites) are commonly added as pastes; it comes as no surprise that wetting and dispersing additives can help optimize the production of such pastes. Bentonites can be activated with either deflocculating or controlled flocculating additives. The question of precisely which product should be employed depends upon the performance features desired in the final system.

10.6.6 The Possible Side-Effects of Wetting and Dispersing Additives

Possible rheology effects (such as the hindrance of sagging and settling) have already been described. The alteration of rheological properties can influence not only the flow behavior, but also the foaming properties. This is why optimally chosen deflocculating wetting and dispersing additives can function as flow and levelling aids, and why they can even help prevent foam formation, thus enhancing defoamer performance. Properly selected additives can even improve anti-corrosion and related properties. Improperly selected wetting and dispersing additives, on the other hand, can impair film formation and anti-corrosion properties. The above examples have been expressly chosen to emphasize that two important factors must simultaneously be considered when selecting the proper wetting and dispersing additive. The first factor is, of course, the additive's influence upon wetting and dispersing properties; the second factor is the summation of the additive's influence upon all the properties exhibited by the formulated coating system.

Guide to Further Reading

Bode, R., Bühler, H., Ferch, H., Statt., B., *Pitture e Vernici* (1973) 8, p. 324
Degussa Corporate Communication, Schriftenreihe Pigmente Nr. 41 (1989) *Aerosil für PVC-Plastisole*, Frankfurt/M., 3. Auflage
Degussa Corporate Communication, Schriftenreihe Pigmente Nr. 12: *Degussa-Kieselsäuren für Siliconkautschuk*, Degussa AG, Frankfurt/M. (1981) 3. Auflage
Dekker, Marcel, Inc., *J. Macromolecular Sci., Chem., Phys., Rev. Macromolecular Chem.*, New York
Eirich, F.R., *Rheology—Theory and Applications* (1969) Academic Press, Inc., New York; Vol. 1 (1956), Vol. 2 (1958), Vol. 3 (1967), Vol. 4 (1967), Vol. 5 (1969)
Lodge, A.S., *Elastic Liquids* (1964) Academic Press, Inc., New York
Mercurio, A., Rheology of Acrylic Paint Resin, *Canadian Paint and Varnish* September (1964)
Patton, T.C., *Paint Flow and Pigment Dispersion* (1965) Interscience Publishers, Inc., New York
Pierce, P.E., Rheology of Coatings, *J. Paint Tech.* (1969) 41 (533), p. 383–395
Reiner, M., *Deformation, Strain and Flow* (1969) Lewis and Co., Ltd., London
Wilkinson, W.L., *Non-Newtonian Fluids* (1960) Pergamon Press, Inc., New York

"Classical" Interfacial Tension Control (Flow, Levelling, and Surface Enhancement)

11 An Introduction to Controlling Interfacial Tension Parameters

11.1 Preface

The growing prevalence of environmentally friendly coating systems has created many new manufacturing and formulating challenges. This introductory chapter, as one of four chapters overviewing interfacial tension parameters, introduces some of the most critical concepts of interfacial tension control. In particular, the types of ingredients and substrates utilized in conjunction with advanced aqueous and high solids coatings systems generally exhibit significantly greater surface tension control challenges than their counterparts in traditional systems. Additives that modify interfacial tension are utilized in paint systems for a variety of reasons. For example, they can be employed as defoamers, flow improvers, anti-cratering aids, or even as substrate wetting enhancers. The common theme in all areas of usage is the reduction or avoidance of surface defects. As shown in Figure 11.1, the exceedingly wide range of surface tension values encountered can span from approximately 20 to 70 dynes/cm. (Please note that, in regard to semantics and nomenclature, 100 dynes/cm = 100 mN/m.)

Many defects in the surface of a paint film can be explained by uncontrolled differences in interfacial tension:

- Poor substrate wetting
- Spray dust sensitivity
- Cratering
- Formation of Bénard cells
- Ghosting, wipe marks
- Poor recoatability
- Air-draft sensitivity

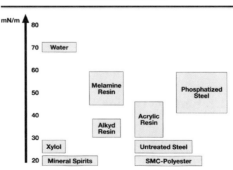

Figure 11.1 Comparative surface tension values

Dimethylpolysiloxane

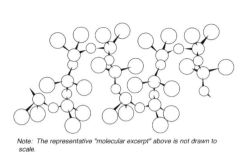

Note: The representative "molecular excerpt" above is not drawn to scale.

Figure 11.2 Three-dimensional molecular networks: Dimethylpolysiloxane

Eliminating these differences will prevent surface and interfacial defects. The surface tension of a coating is a function of the resins and solvents used, and thus can be controlled by selecting the correct raw materials. However, this is only theoretically true, because the practical selection of resins and solvents is normally not based upon their surface tensions, but upon other properties [e.g., durability, hardness, cross-linking mechanism (for resins), volatility, dissolving power, and flash point (for solvents)].

Silicone additives, such as derivatives of the example structure shown in Figure 11.2, are very useful since they are ideally suited for adjusting and controlling interfacial tension; furthermore, they can minimize interfacial tension differences.

11.2 The Influence of Siloxane Additives on Film Properties

Siloxane additives are very versatile and are used in coating systems for many different reasons. (For example, they can be employed as defoamers, anti-blocking agents, flow control agents, and anti-crater additives.) However, in most cases, the effectiveness of these products can be explained by the following three basic features of siloxane additives:

- Surface tension reduction
- Slip improvement
- Controlled incompatibility

11.2.1 Surface Tension Reduction

Due to their surface activity, silicones concentrate at the paint surface. (Certain exceptions to this "rule" will, however, be discussed later.) One of their most

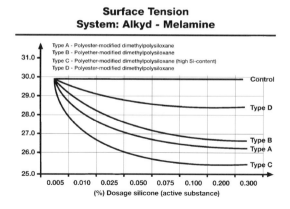

Figure 11.3 Surface tension as contingent upon chemical structure

characteristic properties is their ability to reduce surface tension. Depending on the chemical structure, the degree of reduction varies, and can usually be measured conveniently with the ring detachment method (more detail about this later).

After application onto a substrate, the paint composition changes due to solvent evaporation and/or curing reactions. This leads also to a possible change in surface tension. With (non-volatile) silicones, it is not only possible to reduce the surface tension, but it is also possible to stabilize the surface tension at a low level throughout the entire drying process. Examples of surface tension—as contingent upon chemical structure—in a representative system are shown in Figure 11.3. (Please note that virtually all volatile media—including water—can be thought of as belonging to the "solvent realm". Although certain interfacial phenomena do exhibit rather unique behavior in water-based systems, at least half of all silicones are designed expressly for usage in such systems. Much of this textbook's focus is placed on water-based and related environmentally friendly coatings.)

11.2.2 Slip Improvement

In addition to controlling surface tension, certain silicones also influence the surface slip of a coating. Slip (as shown in Figure 11.4) is measured by monitoring the force necessary to draw a defined weight across the coating. Silicone additives generally reduce friction, i.e., they improve slip, depending on their chemical structures and concentrations. Better slip may actually be the desired property itself, but oftentimes slip is desirable for secondary reasons, especially since coatings providing improved slip may simultaneously display better mar and scratch resistance, along with improved anti-blocking properties. An added benefit in nearly all cases is less susceptibility to soiling.

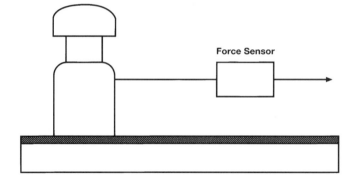

Figure 11.4 Slip measurement

11.2.3 Controlled Incompatibility

Depending on their chemical structures, silicone additives are more or less compatible with the binder solution; compatibility features are primarily controlled by the molecular weight of the additive and the presence of modifying side-chains. For reference purposes, additional "rules of thumb"—regarding compatibility—are shown below:

- Extremely incompatible silicones will lead to surface defects (like craters) and are generally used only for hammertone coatings.
- "Compatible" silicones will reduce surface tension and improve surface slip—as described previously—without creating surface defects. (One may also choose to "mix or match" surface tension and surface slip properties, so to speak. For instance, certain exceptional silicones—as described later—can be designed to reduce surface tension without affecting slip.)
- Silicones which are just on the borderline between compatibility and incompatibility ("controlled incompatibility") function as defoamers.

From the coating formulator's perspective, one must design dependable, defect-free alternatives both for reducing surface tension and for optimizing performance; as a result, newly developed silicone chemistries have been introduced to provide a high performance, cost-effective avenue of approach.

Within the context of performance optimization, the following topics will be discussed:

- The interdependence of surface tension and substrate wetting
- Crater formation and elimination
- Bénard cells, flooding/floating, flow, and air-draft sensitivity
- Surface slip
- Foam removal

During the paint's transition from a liquid, extremely mobile state to a dry immobile film, many system-dependent differences in surface tension occur. One prime source of such differences—and also of the resultant defects—arises within the coating system itself (solvent evaporation and/or the cross-linking reaction of the resin). Another source arises from external causes (overspray, dust particles, or contamination of the substrate). Through either the reduction of surface tension or through the avoidance of surface tension differences, silicone additives can successfully combat the defects mentioned at the beginning of this chapter. Moreover, such additives can additionally improve the following properties of the dry paint film:

- Slip
- Scratch resistance
- Anti-blocking

Through controlled incompatibility, such additives can furthermore function as defoamers during paint production and application.

11.3 New Chemistry for Environmentally Friendly Systems

Polysiloxanes (also abbreviated "silicone additives" or "silicones", sometimes even as "siloxanes") can often be employed without an understanding of the underlying chemistry; however, it is still helpful to outline and understand a few of the more basic principles of silicone technology. Armed with the basic principles presented in this chapter, one can more easily begin to identify and understand the correlation between structural features and resultant end-use properties.

11.3.1 Polydimethylsiloxanes

Most silicone additives are derived from the basic structure of polydimethylsiloxane. Through variations in the chain length, a broad palette of products and derivatives with great variation in properties and features can be produced; examples of the myriad performance possibilities in regard to flow control, anti-floating, surface slip, and defoaming are shown in Figure 11.5.

Short-chain silicone additives are relatively compatible in most paint systems. Typical silicone properties (such as low surface tension, improved flow, etc.) are exhibited. On the other hand, long-chain molecules are rather incompatible and can potentially lead to craters and/or hammertone effects. Accordingly, pure poly-dimethylsiloxanes (silicone oils) are hardly ever used in modern paints.

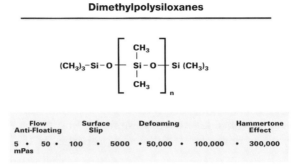

Figure 11.5 Performance possibilities in regard to flow control, anti-floating, slip, and defoaming

11.3.2 Polyether Modification

A more elegant method than chain-length-dependent compatibility consists of modifying the silicone backbone structure with side-chains. Most silicone additives employed today are "organically modified polysiloxanes" (Fig. 11.6).

Generally, the most important feature is the modification with polyether side-chains; nevertheless, one should always utilize pendant carbon groups as the linkage between the polyether side-chains and the alternating silicon/oxygen backbone structure. In contrast, pendant oxygen group linkages attached directly to the backbone (as shown in Figure 11.7) should always be avoided. Consequently, one must be careful to select silicone additives from the appropriate supplier.

Through the introduction of various types and numbers of side-chains, compatibility can be improved or modified. The relationship or proportion of dimethyl

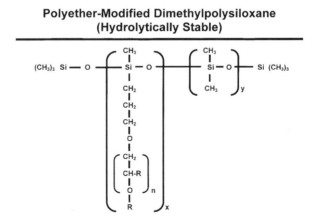

Figure 11.6 Polyether-modified moiety

**Polyether-Modified Dimethylpolysiloxane
(Hydrolytically Unstable)**

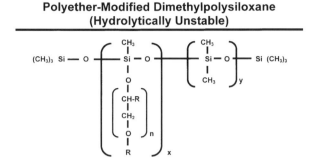

Figure 11.7 Pendant oxygen groups in hydrolytically unstable molecule

groups to polyether modifications (Y to X) allows one to control the degree of compatibility. Simultaneously, this relationship exerts a direct influence upon surface tension (the greater the number of dimethyl groups available, the lower the surface tension, as a rule of thumb). Furthermore, the structures of the polyether chains themselves can be varied. A very important variable here is the polarity of the structural elements. The polyether chain itself is composed of ethylene oxide (EO) and may also include propylene oxide (PO) units. (The synthetic possibilities are endless; only a few are covered here.) Polyethylene oxide is very hydrophilic (polar); however, polypropylene oxide is comparatively hydrophobic (non-polar). Through utilizing various proportions of EO to PO, one can control or modify the polarity of the entire silicone additive. For example, a higher proportion of EO raises the polarity so that the additive is water-soluble and thus more compatible in very polar systems; however, the tendency toward foam stabilization increases. On the other hand, a higher proportion of propylene oxide (PO) reduces the water solubility, thereby minimizing foam stabilization.

11.3.3 Polyether-Modified Methylalkylsiloxanes

Another possibility for modifying the silicone structure consists of replacing (either partly or entirely) the methyl groups of the dimethyl structures with long-chain alkyl groups. Polyether-modified methylalkylsiloxanes, as shown in Figure 11.8, are the result.

In comparison to their polydimethylsiloxane counterparts, such products clearly demonstrate higher surface tensions, and have less influence on surface slip properties. They are often found as active constituents in defoamers. Polymethylalkylsiloxanes can, of course, be organically modified with polyether chains in the same fashion as polydimethylsiloxanes. As shown in Figure 11.9, the resultant surface tension value of any particular polyether-modified methylalkylsiloxane is, in part, dependent upon the exact length of the alkyl chain.

Polyether-Modified Methylalkylpolysiloxane

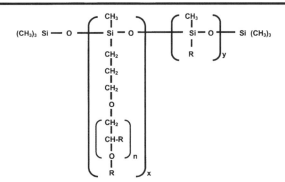

Figure 11.8 Methylalkyl moiety

Surface Tension of Methylalkylpolysiloxanes

- R	Surface Tension [mN/m]
- CH$_3$	20.6
- CH$_2$CH$_3$	26.2
- (CH$_2$)$_2$CH$_3$	26.2
- (CH$_2$)$_3$CH$_3$	27.6
- (CH$_2$)$_4$CH$_3$	28.3
- (CH$_2$)$_5$CH$_3$	28.2
- (CH$_2$)$_7$CH$_3$	30.4
- (CH$_2$)$_9$CH$_3$	31.4
- (CH$_2$)$_{11}$CH$_3$	32.5
- (CH$_2$)$_{13}$CH$_3$	33.5

Figure 11.9 Comparative surface tension values as correlated to molecular structure

11.3.4 Other Modifications

Polyether-modified polysiloxanes are temperature stable up to approximately 150°C (300°F). At higher temperatures, the polyether chains themselves decompose; however, through the utilization of non-polyether structural elements, various thermally stable products can be produced. For this express purpose, one can employ special polyester and aralkyl groups. The correspondingly modified polysiloxanes are thermally stable up to 220°C (430°F) (or even higher temperatures) and can therefore be employed at significantly higher oven and cure temperatures than the traditional polyether moieties.

11.3.4.1 Reactive Silicones

As a rule, silicone additives are not reactive; this means that they do not take part in the cross-linking reaction of the resin itself. However, for special application areas, it may be desirable for the silicone additive itself to be incorporated into the binder structure. In such situations, reactive silicone additives with special "end group" structures are available. Reactive products possess, for instance, either primary hydroxyl groups (for reaction with isocyanates) or double bonds (for reaction with certain UV curable systems) at the end of the organic modification.

11.3.4.2 Silicone Surfactants

Chemically speaking, "silicone surfactants" can be classified as polyether-modified polydimethylsiloxanes; nevertheless, their molecular weights are considerably lower than those usually encountered with other silicone additives. (Please note that many disparate, context-based interpretations of the word "surfactant"—all equally correct and acceptable—exist in the industry. Additional nomenclature variants, not just for the word "surfactant", are in common use.) A typical silicone surfactant backbone consists of only a few Si–O units and contains, on average, merely one polyether chain. Because of this, such products have a very pronounced surfactant structure (intermeshing of both polar and non-polar entities; more information is provided in subsequent sections of Part III), and therefore tend to lower the surface tension to levels often matching those of fluorosurfactants—without simultaneously increasing the slip properties of the coating system. Fluorosurfactants have long been employed, especially in aqueous coatings, to reduce surface tension to dramatic levels; however, such fluoro products often suffer from the disadvantage of increased foam. In contrast, silicone surfactants do not stabilize foam. (Please note that the following terms—among others—are often used synonomously in various quarters of the industry: "fluorosurfactants", "fluoro products", "fluorine derivatives", "fluorocarbon surfactants", etc.)

If increased slip is required, silicone surfactants can obviously be combined with other silicone additives (such as longer-chain polyether-modified polysiloxanes).

11.4 Synopsis of Chemical Structure Parameters

Many recent silicone chemistry advances have occurred; a synopsis of the performance attributes associated with selected structural elements is shown in Figure 11.10.

Structure-Performance Relationships

• Organic Modification of the Polysiloxane

$$\begin{array}{l} -\ \underset{|}{\overset{|}{Si}} - O - R \quad [\text{ prone to hydrolysis }] \\ \\ -\ \underset{|}{\overset{|}{Si}} - R - \quad [\text{ resistant to hydrolysis }] \end{array}$$

• Adjusting the Compatibility

A. Thermal stability up to 150° C/ 300° F (polyether can be utilized)

$$-\left[CH_2 - \underset{\underset{R_2}{|}}{CH} - O\right]_n$$

B. Thermal stability at more than 150° C/ 300° F requires non-polyether (for instance, polyester or aralkyl)

C. Polyester example

$$\left\{O - \overset{\overset{O}{\|}}{C} - R - \overset{\overset{O}{\|}}{C} - O - R\right\}_n$$

• Reactivity

A. No reactivity – good intercoat adhesion

B. Reactivity – bad intercoat adhesion

• Alkyl group length

CH₃		(CH₂)ₓ – CH₃
High	"Slip"	Low
Low	Surface Tension	High

Figure 11.10 Synopsis of performance attributes: Structure/performance relationships

Guide to Further Reading

Adamson, A.W., *Physical Chemistry of Surfaces* (1967) Interscience Publishers, Inc., New York

Bikerman, J.J., *Surface Activity*, Second Edition (1967) Academic Press, Inc., New York

Bikerman, J.J., *Surface Chemistry*, Second Edition (1958) Academic Press, Inc., New York

Flood, E.A. (Ed.) *Solid-Gas Interface* (1967) Marcel Dekker, Inc., New York

Gregg, S.J., *Surface Chemistry of Solids*, Second Edition (1961) Chapman and Hall, London

Harkins, W.D., *Physical Chemistry of Surface Films* (1952) Reinhold Publishing Co., New York

Haubennestel, K., Bubat, A., *Silicones in Waterbased Paints*, XIX Congress AFTPV (1991) Nice, pp. 257-260

Mysels, K.J., *Introduction to Colloid Chemistry* (1959) Interscience Publishers, Inc., New York

Orr, E.W., *Silicones in the Coatings Industry* (1993) Lecture and monograph presented at California Polytechnic State University

Shaw, D.J., *Introduction to Colloid and Surface Chemistry* (1966) Butterworths Scientific Publications, London

12 The Interdependence of Surface Tension, Substrate Wetting and Other Surface Phenomena

12.1 Substrate Wetting

Substrate wetting depends primarily upon the surface tension of the paint and upon the critical surface tension of the substrate to be coated. As a general rule, the surface tension of the paint must be lower than or equal to the surface tension of the substrate. Improper wetting, such as "crawling" or "beading-up" of the paint, is to be expected when the surface tension of the paint is above the "critical surface tension" of the substrate. In order to demonstrate the importance of the aforementioned critical surface tension aspect, a case study was performed in two coating systems; the results are displayed in Tables 12.1 to 12.2.

Table 12.1 Critical Surface Tension: Case Study (Comparison of Control to Silicone-Containing Counterpart)

Paint system	γ_c (mN/m) without silicone	γ_c (mN/m) 0.1% Dimethylpolysiloxane copolymer
Alkyd/melamine	35	24
2-Pack polyurethane	32	25

Table 12.2 Critical Surface Tension: Case Study (Including Paint Films after "Solvent Wipe" or Surface Repair)

Paint system	γ_c (mN/m) without silicone	γ_c (mN/m) without silicone after "solvent wipe"	γ_c (mN/m) 0.1% Dimethyl-polysiloxane copolymer	γ_c (mN/m) 0.1% Dimethyl-polysiloxane copolymer after "solvent wipe"
Alkyd/melamine	35	34	24	32
2-Pack polyurethane	32	32	25	31

12.1.1 Case Study (Before Processing)

Table 12.1 compares the control to the silicone-containing counterpart. Of course, significant surface tension reductions occurred—after silicone utilization—in both paint systems, thus making it possible for these particular systems to provide proper wetting on any substrates demonstrating surface tensions of more than 24 mN/m or 25 mN/m (depending upon exactly which paint system of the exhibited pair was employed).

12.1.2 Case Study (After Processing)

Table 12.2 examines what might happen to critical surface tension when, for instance, the first layer of a multi-layer system is allowed to dry, inadvertently wiped (only in certain areas) with solvent and/or otherwise "processed" or repaired in anticipation of topcoat application. As evident in the increased "post-repair" surface tension of layer number one (now considered the "substrate" in regard to the relevant surface tension calculations involved with the topcoat)—the critical surface tension value in the wiped/repaired areas is either 32 or 31 (depending upon the system). Keeping in mind that the unwiped/unrepaired areas have respective surface tensions of 24 or 25, then a heterogeneous surface tension exists. In practice, this may cause improper surface features (such as ghosting) because of the propensity of the liquid paint layer to continue flowing toward the areas of higher surface tension.

Substrates with rather low surface tensions (for example: plastics, surfaces with oil residues, and certain contaminated metals) are not easily wetted. Aqueous paint systems, because of their water content, are obviously higher in surface tension than paint systems with organic solvents, and demonstrate, in many cases, problems with substrate wetting.

Silicone additives reduce the surface tension of the paint and accordingly lead to more optimal substrate wetting. Of special interest here are silicones containing dimethyl structures, since such structures are capable of efficiently reducing surface tension. As mentioned previously, silicone surfactants are also particularly suitable for *aqueous* systems.

12.2 Craters: Formation and Elimination

The causes for crater formation can be quite diverse. Craters may be created, for example, by overspray falling into a freshly sprayed paint layer (a layer that is still mobile). Fine spray droplets from the overspray can lead to craters when the surface tension of such droplets is lower than the surface tension of the still mobile paint surface (Fig. 12.1). Surface tension differences of 1 to 2 mN/m are sufficiently large

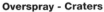

Overspray - Craters

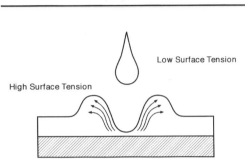

Figure 12.1 The physics of crater formation

enough to cause spreading of the droplets, thus leading to craters. In cases where the surface tensions of both materials are the same, or if the spray droplets have a higher surface tension, then there is no spreading action of the droplets and consequently no crater formation. Small dust particles falling into the still-mobile paint layer can also produce the same effect as overspray droplets.

Craters can sometimes originate from unclean or contaminated substrates, for example, from fingerprint-contaminated panels. When such substrates display inordinately low surface tensions, they will tend to cause craters whenever subsequent paint layers are applied. Through proper utilization of a silicone additive, surface tension is lowered and therefore the paint becomes visibly less susceptible to disturbances or disruptions—whether or not such disturbances arise from the outside environment (overspray, dust particles, etc.), from the substrate (contamination), or even from within the film itself (gelled particles). Because the most effective surface tension lowering agents are required, then polydimethylsiloxanes (modified, of course, as appropriate) are often employed as the anti-crater additives of choice. (Please note that, as a matter of industry convention, the word "modified" does not always appear in the generic designations of "polysiloxanes" and/or related products that have indeed been modified.)

12.3 Bénard Cells, Flow, Flooding, and Air-Draft Sensitivity

12.3.1 Bénard Cell Formation

Because of the processes involved with film curing and resin drying, turbulent flow patterns often arise in coating systems. Such flow patterns can be constant in nature, methodically transporting material from the lower layers of the drying film up to the very surface of the film. This sometimes leads to the formation of specialized cell structures denoted as "Bénard cells". (Of course, this term has been introduced before;

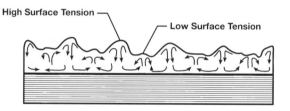

Figure 12.2 Turbulence, improper flow, and surfaces that are not uniformly smooth

however, various chapters cover selected aspects of this phenomenon within different contextual frameworks. Certain elements of review are necessary to set the stage for interweaving "new" concepts into the "old"—or for introducing new material in either different and/or expanded frameworks.) Differences in density, temperature—and especially surface tension—often constitute one of the prime driving forces behind this cell formation process.

The solvents evaporating from paint carry a low viscosity medium of lower density from the deeper zone to the surface. The medium spreads out, the solvents evaporate, and the system becomes more viscous. It acquires a higher density and sinks back into the bulk of the paint which is richer in solvent. The pigments are carried along and whirl about in continuous eddies. As the paint dries and becomes more viscous, the movement gradually comes to a halt. The structures that are seen on horizontal surfaces are *called Bénard cells*, but on vertical surfaces the deformation of these cells produces *silking*. (Please note that there are actually several dozen factors contributing to Bénard cell formation; only selected variables and constraints are mentioned here.)

The turbulence of the Bénard cells also leads, in many cases, to improper flow properties and consequently to paint surfaces that are not uniformly smooth (Fig. 12.2). Silicone additives can combat cell formation by orienting to the surface (in this case, the air/liquid interface). They form a micro layer (Fig. 12.3) which effectively immobilizes the surface region. As a result, there are no zones of differential surface tension and consequently no Bénard cells.

12.3.2 Flow (Proper and Improper)

A liquid tends to wet areas of higher surface tension. One way of expressing this is to say that the more "active" system spreads on the more "inactive" one. Formulators and applicators alike can apply this knowledge to explaining how craters may form during a spray operation. For instance, if the spray dust has a lower surface tension than the already applied liquid paint, then craters occur. This is because the spreading direction is from areas of lower surface tension to areas of higher surface tension. By pushing forward the already applied paint material, a crater is formed.

On the other hand, if the overspray material has higher surface tension—then such material will not cause craters. It will stay on top of the already applied paint. (Of

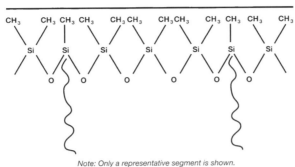

Orientation to the Interface

Note: Only a representative segment is shown.

Figure 12.3 Micro layer (chemical orientation)

course, the force of gravity may indeed "flatten" the material to some extent, but surface-tension-induced crater formation will not result. Extraneous variables, such as chemical-induced spreading, may also play a role in a very small percentage of cases.)

12.3.3 Flooding

In a pigment-containing paint system, the pigments themselves will naturally participate in the circular flow patterns. This applies not only *on* the surface, but also *beneath* the surface. (In regard to definitional distinctions, some authors may indeed denote on-the-surface problems as "floating"—and beneath-the-surface problems as "flooding". The nomenclature of the terms is such that "floating" is often considered a subset of "flooding". But words in the coatings industry—as in so many other industries—are not necessarily standardized; many authors quite commonly modify or even reverse the meanings. Context surely makes a difference, and therefore this explains many of the current "word debates" in the industry. These issues are mentioned only so that they do not become stumbling blocks to our continued study. Fortunately, we scientists live in a three-dimensional world often populated by rather easily observed physical phenomena, so the ethereal realms of "semantical distractions" can indeed take a backseat to the more pressing real-world issues of performance enhancement.) Especially in cases where different pigments with correspondingly different mobilities interact, pigments can become widely separated because of improper flow patterns and other related film disturbances. On the paint surface, pigments are no longer homogeneously distributed. On typical horizontal surfaces, Bénard cell formation occurs; and on vertical surfaces, silking occurs. Oftentimes, a rather pronounced deformation of the surface texture can manifest itself. The coating displays non-optimal flow; a special term describing one variety of this defect is "orange peel". (Chapter 6 includes augmented information about flooding, floating, and certain related phenomena; in addition, the remainder of Part III contains selected mention of such phenomena where appropriate.)

12.3.4 Air-Draft Sensitivity

Dependent upon the drying conditions and also upon the particular features of the solvent mixture, the drying of the resin solution can occasionally become so strongly influenced by surface defects that the entire top surface of the film is completely disrupted. This extreme effect, especially pronounced in furniture coatings, is denoted as "air-draft sensitivity" and is generally caused by air-draft conditions that place extreme stress on the top layers of the coating. Polysiloxanes can be employed to control or eliminate this deleterious effect.

Guide to Further Reading

Fisk, P.M., *Physical Chemistry of Paints* (1965) Chemical Publishing Co., Inc., New York

Lenz, R.W., *Organic Chemistry of Synthetic High Polymers* (1967) Interscience Publishers, Inc., New York

Martin, R.W., *Chemistry of Phenolic Resins* (1956) John Wiley and Sons, Inc., New York

Pratt, L.S., *Chemistry and Physics of Organic Pigments* (1947) John Wiley and Sons, Inc., New York

Reid, R.C., Prausnitz, J.M., Sherwood, T. K., *The Properties of Gases and Liquids* (1977) McGraw-Hill Book Company, New York

Solomon, P., *Chemistry of Organic Film Formers* (1967) John Wiley and Sons, Inc., New York

Turner, G.P., *Introduction to Paint Chemistry* (1967) Chapman and Hall, London

13   Advanced Methods of Interfacial Tension Control

13.1 Cell Structures and SFCA

Through the proper utilization of a silicone additive, coating surface tension is stabilized at a low level; therefore, surface tension differences cannot develop within the film. As a logical corollary (especially since such surface tension differences are integrally connected with the formation of cell structures), then the subsequent elimination of surface tension gradients causes the disappearance of many undesirable effects. Many different silicone additives can be used to prevent Bénard cells. However, in the case of flooding, one must be aware of the fact that the silicone influences only superficial cell structure itself, rather than underlying causes such as the different intracoating, beneath-the-surface pigment mobilities. It is therefore essential to also employ specialized wetting and dispersing additives to influence such pigment mobilities, thereby allowing the silicone itself to function as a supporting element of the effectively controlled system.

Through the introduction of the silicone as a surface flow control additive (SFCA), one can properly influence *the type of surface structure formed*. For example, low surface tensions lead to short wave structures on the top of the film; while on the other hand, high surface tensions lead to comparatively long wave structures.

To avoid the surface flow problems associated with air-draft sensitivity, products demonstrating dramatic surface tension reduction properties are most effective. Interestingly enough, the rheology of the coating also exerts a very strong influence upon surface flow. This means that surface flow can even be influenced by special rheology modifiers and by wetting and dispersing additives. (Please note that, in certain circles of the industry, the terms "flow" and "levelling" tend to be used interchangeably; nevertheless, there often exist rather noticeable differentiating characteristics—as described later in Chapter 15.)

13.2   Special Factors Involving Surface Slip and/or Foam Removal

As mentioned briefly before, certain silicone additives can obviously improve the slip properties of coating surfaces. Oftentimes, it is not merely slip itself, but rather the close association of slip properties with other desirable coating features that is of prime interest. Surfaces with better slip are generally more scratch resistant, less easily soiled, easier to clean, and more resistant to blocking. The degree of surface slip improvement is dependent primarily upon the chemical structure of the silicone and, in particular, upon the proportion of dimethyl groups present. Products with many dimethyl groups result in dramatic slip improvement properties, whereas methylalkylpolysiloxanes result in visibly less improvement.

Silicone "surfactants", however, because of their short backbone chain lengths, will not yield significant slip improvement in most paint systems. If, for whatever reason, higher slip values are required, silicone surfactants must be combined with other silicone additives. (The measurement of slip properties is described in a later section.)

Foam can, on one hand, be caused by improper usage of silicone additives. On the other hand, the judicious usage of silicone additives can actually result in the removal of foam. Of fundamental importance in distinguishing between proper and improper performance are both the polarity and the compatibility of the candidate silicone additive. Highly compatible silicone additives with lower surface tensions tend to display higher probabilities of stabilizing foam. In special cases where corresponding products (of high compatibility) lead to foam problems, one should instead employ products with higher surface tensions (for example, methylalkyl-modified products). Such products are not classical defoamers per se; nevertheless, they can be considered as "silicone flow additives with defoaming properties." In cases where pure defoamer products must be employed, one might occasionally evaluate even more incompatible products (such as unmodified methylalkylpolysiloxanes).

13.3 Synopsis: The Correlation of Ten Performance Parameters with Surface Tension Reduction

Thus far, a broad spectrum of performance parameters (including substrate wetting, craters, Bénard cells, flooding, flow, air draft sensitivity, slip, and foam) has been discussed. All of the above parameters are, of course, influenced by surface tension; accordingly, *a synopsis of additive selection suggestions* and performance improvement methodologies, as related to surface tension reduction, is shown in Tables 13.1 to 13.3.*

13.4 Intercoat Adhesion

Silicones, as a general class of products, have an unearned reputation in some corners of the industry for being unrecoatable and for deleteriously affecting intercoat

---

*The interfacial tension principles presented above apply, for the most part, to nearly all coating systems; nevertheless, some specialized formulations (especially those in the UV/EB arena) may require special attention. Accordingly, Appendix IV provides a broad-based introduction to additive usage in certain specialized formulations.

Table 13.1 Silicones: Guide to Additive Selection (Part I)

Desired property	Basics	Surface tension reduction
13.3.1 Improved flow/levelling	High surface tension = high energy to reduce surface area = good flow/levelling	more ——————→ less (rising)
13.3.2 Prevention of foam stabilization	Strong surface tension reduction by compatible silicones leads to foam stabilization	more ——————→ less (falling)
13.3.3 Prevention of ghosting	Occurs when coating is applied onto another dry paint film that does not have uniform surface tension (silicone partly removed by sanding and/or cleaning with solvent). Areas of different surface tension in the first layer induce material transport in the second layer. Visible difference in film thickness	more ——————→ less (rising)
13.3.4 Improved sag resistance	(Please see the somewhat similar surface tension situation described in the "Improved substrate wetting" subsection of this diagram series)	more ←—————— less (rising toward more)

Table 13.2 Silicones: Guide to Additive Selection (Part II)

Desired property	Basics	Surface tension reduction
13.3.5 Reduced air-draft sensitivity	Rapid evaporation due to air movement leads to strong eddy currents and to a complete rupture of the film	more ——___ less
13.3.6 No influence on "aluminum-orientation"	A change of surface tension automatically leads to different orientation of the aluminum flakes	more ——————→ less (falling)
13.3.7 Improved anti-cratering/ overspray acceptance	Lower surface tension is needed to avoid craters caused by particles or contaminants (i.e., dust) that have fallen into a wet paint film	more ←—————— less

Table 13.3 Silicones: Guide to Additive Selection (Part III)

Desired property	Basics	Surface tension reduction
13.3.8 Prevention of Bénard cells	During the drying process, solvent evaporation leads to a non-uniform surface tension on the paint surface. Eddy currents are formed. Different pigment mobilities lead to pigment separation. Note: Bénard cells also develop in unpigmented systems	more less
13.3.9 Improved slip	Large number of dimethyl-groups = high surface tension reduction = high slip	more less
13.3.10 Improved substrate wetting	To wet a substrate, the surface tension of the wet paint must be equal to or lower than the surface tension of the substrate. Hint: Lower and/or more Newtonian rheology of the paint supports poor substrate wetting	more less

adhesion. However, through the proper choice and usage of silicone additives, virtually all problems can be eliminated. Silicone additives migrate to the paint surface because of their surface activity. Since they, as a general rule, have no reactive groups, they will not participate in the drying/curing mechanism of the binder. In other words, this means that silicones remain actively mobile throughout the entire curing process. This is demonstrated, for example, by the fact that one can wipe off or remove (utilizing solvent) the additive from the surface.

When recoating of the silicone-containing layer occurs, the silicone *does not remain in the surface of the first layer* (namely, the *interfacial boundary zone* between the two paint films). As anticipated, the silicone migrates—because of its mobility and surface activity—into the new surface, the second paint layer. Between the two paint films, no silicone remains and, accordingly, the intercoat adhesion is not negatively influenced (Fig. 13.1). The ability of the silicone to migrate is partly a function of its molecular weight and structure.

However, two factors can negatively influence intercoat adhesion: (1) overdosage of the silicone additive, and (2) baking temperature of the coating. For each resin/silicone combination, there exists an optimal concentration of the silicone additive, above which there is no significant additional benefit. The usage of higher concentrations can induce negative side effects such as poor intercoat adhesion. When high usage levels of silicone are present in the first paint layer, most of the additive will migrate into the second layer upon overcoating. Nevertheless, some molecules may remain in the

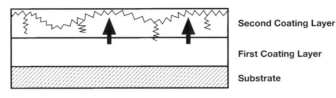

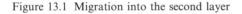

Figure 13.1 Migration into the second layer

interface between the two layers and reduce the intercoat adhesion, so it is essential to first determine the necessary additive dosage in a systematic test series or in a "ladder-study". Then, after this initial determination, one should employ only the minimal concentration required. For every resin/silicone combination, there is an optimal silicone usage level. Higher usage levels will not generally result in further advantages (wetting, anti-cratering, and/or slip properties).

In addition to dosage parameters, baking variables—including temperature—can also influence intercoat adhesion. Consider, for instance, systems in which the proper baking temperature of the first layer (containing the silicone) is exceeded. What happens? The adhesion between the two layers may be noticeably reduced. This deleterious effect occurs when the polyether chains of the additives decompose at higher temperatures [140 to150°C (280 to 300°F)] because of oxidation. (Please note that a discussion of baking variables here certainly does not imply that all polysiloxanes in Part III are designed for baking systems.)

While the silicone backbone itself is stable, the polyether moiety decomposes step by step in an induced oxidation process. This thermal degradation can be explained as a radical chain reaction mechanism during which intermediate reactive sites are formed in the polyether chain (e.g., COOH groups); and during which there is the possibility that the silicone molecules may react with the resin molecules via these groups. The end result is that the silicone in the paint film loses its mobility, and an insoluble "release layer" is formed. If such an overbaked paint film (Layer #1) is now overcoated (with Layer #2), the silicone can no longer migrate through the second layer. It will remain in the interface between the two layers and impair intercoat adhesion.

Through this oxidation process, reactive groups are created so that the silicone additive now becomes an integral part of Layer #1, and therefore loses its migration ability. The resultant decomposition products tenaciously remain between the two layers and adversely affect adhesion by acting as an irrevocable release layer (Fig. 13.2). (Of course, manufacturers in the label industry may indeed be looking for this type of behavior in either baked and/or non-baked systems. Performance attributes can be completely tailored, but it is important to mention that scores of more efficient "release layer methods" can be designed. Within the perspective of most coating manufacturers, however, improper intercoat adhesion is obviously nothing more than a nightmare to be avoided.) Since the thermal instability mentioned above is caused by the presence of polyether chains, it is therefore possible (by exchanging the polyether

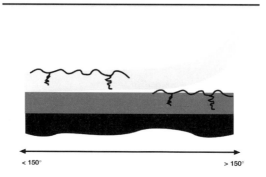

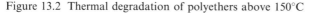

Figure 13.2 Thermal degradation of polyethers above 150°C

chains with more stable groups) to avoid this effect. For example, certain silicone additives with polyester or aralkyl modifications remain stable at temperatures up to 220–250°C (430–480°F).

13.5 How to Measure Surface Tension and Slip

Because silicone additives influence surface tension and slip, it is obviously important to measure the resultant properties in a scientific manner. For instance, in order to determine the surface tension of liquids, the ring detachment method can be employed. Examples of this method, along with other useful techniques, are shown in Fig. 13.3.

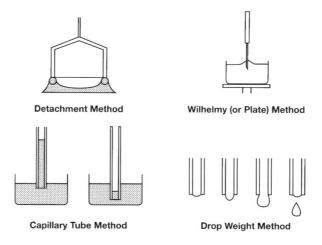

Figure 13.3 Surface tension measurement methodologies

The precise determination of the critical surface tension of solid substrates may require the utilization of the comparatively expensive and time-consuming contact angle method. In contrast, though, the rather rapid and inexpensive determination of approximate values can be achieved merely by utilizing specialized test inks (as described in ASTM D 2578/67).

In order to investigate the slip features of coatings, a specially designed weight can be placed upon the dried coating and then gradually pulled across the surface. The amount of force required to pull the weighted body across the planar surface can be directly correlated to slip.

13.6 Special Challenges: "Ghosting"—Wipe Marks

The special phenomenon of "ghosting" (or wipe marks) occurs when one coating is applied onto another paint film that does not have a uniform surface tension. The areas of differential surface tension (high and low) in the first paint layer will induce material transport in the second layer (from the area of lower surface tension towards the area of higher surface tension). This macroscopic movement of paint material eventually leads to differences in film thickness which are visible on the coating surface.

This situation normally occurs when the first coating contains a silicone and, for some reason, has to be partially sanded (e.g., to remove dust particles or other defects.) After sanding, the surface is then cleaned with solvent. The combined sanding and cleaning operations will remove the silicone (or at least part of it) from the paint surface and thus create surface tension differences.

This defect will be very obvious when the surface tension difference is very pronounced. In contrast, slight differences will not be visible; therefore, this observation—in and of itself—can actually indicate one rather effective method of possibly preventing the problem. The silicone in the first paint layer should have a fairly high surface tension (close to that of the resin system). If such a silicone is used in the first layer, then removal of the silicone by sanding/cleaning will not create a significant difference in surface tension. Accordingly, the ghosting effect will not be visible. Methylalkyl polysiloxanes are often the preferred products for use under such conditions.

13.7 Special Challenges: "Ghosting", Wetting, and Other Phenomena

In Fig. 13.4, a case study was employed to simulate ghosting. First of all, the substrate was coated with a paint in which a polysiloxane was used. Afterwards, this first layer

Ghosting

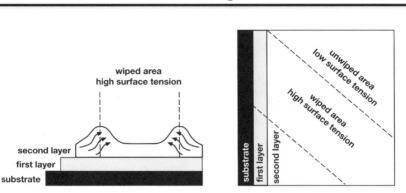

Figure 13.4 Ghosting

was wiped with a solvent-containing cloth. Then the differential surface tensions of the wiped and the unwiped surfaces were determined by contact angle measurement. The surface tension of the unwiped area was 25 mN/m, that of the wiped area 35 mN/m. Then the surfaces were recoated with the same paint system. (In such studies, it is not necessary, however, that the same paint system be used. Ghosting phenomena may also appear, for instance, when certain primers and topcoats are used in conjunction with one another.) In the experiment at hand, wipe marks became clearly apparent in the second layer. As a rule, ghosting involves residual images which can be observed in the perimeter of the wiped area, precisely where beads and/or slightly raised structures are formed.

When a coating is applied over a substrate, it should easily wet the substrate. Substrate wetting is controlled mainly by the surface tension of the liquid paint material and the solid substrate. Generally speaking, for the achievement of proper substrate wetting, it is necessary for the wet paint to have a lower surface tension than the substrate. Poor wetting or perhaps even crawling will occur if the paint's surface tension is higher than the substrate's. Reducing the surface tension by means of silicone additives can help ensure good wetting.

In some cases, the wetting is perfect in the time period directly after application of the coating. However, due to the evaporation of solvents during the drying phase, the surface tension of the paint film increases and can become higher than the surface tension of the substrate. If this state is reached while the paint film is still very liquid and mobile, then crawling will occur. Also, in this case, a tailored silicone can help alleviate the problem, because the silicone additive (non-volatile) stays in the paint film throughout the entire drying process and maintains the surface tension at an appropriately low level. Other factors may, of course, influence wetting (e.g., viscosity and film thickness) and should not be neglected. If a paint of low viscosity exhibits wetting problems in a thin film, such problems may not be visible when viscosity and/ or film build are increased.

Especially in water-based systems, substrate wetting can be a very serious problem because of the high surface tension of water. Conventional silicone additives often cannot prevent wetting defects since they do not adequately reduce the surface tension. The correct products to use under such conditions are often silicone surfactants: such products provide very low surface tension values in waterborne coatings, and therefore avoid wetting problems. They can replace fluorosurfactants which work similarly, but which generally demonstrate a pronounced tendency to foam (fluoro products are also more expensive).

In general, coating materials with low surface tensions are less sensitive to nearly all avenues of contamination from either the substrate or from outside sources; nevertheless, one must ensure that surface tension differentials—such as those occurring with Bénard cells—do not occur. What would happen, for instance, if uncontrolled differentials were to occur at the air/liquid interface? Consider the case of Bénard cells: after application of the coating, solvents evaporate from the film, thus leading to the previously discussed change in surface tension. However, this change is not absolutely uniform across the entire paint surface. In one area, more solvent has already escaped the paint film, and the surface tension has increased accordingly. In another area, the solvent content is still higher, and therefore the surface tension is lower. Thus, surface tension differences develop across the surface. (The dozens of additional "Bénard-cell-instigating and/or -mitigating factors" are not discussed here.)

Such differences may be rather minuscule, and one may even be tempted to disregard them; however, as mentioned in a previous chapter, these differences initiate movement of paint material from lower to higher surface tension regions. Accordingly, eddy currents are formed all over the paint surface. Differences in temperature and density (which can sometimes result, in part, from solvent evaporation) often accentuate this effect. These localized currents can develop in clear coatings as well as in pigmented systems. Quite unsurprisingly, they can be highly pronounced if the paint viscosity is low and the film thickness is high.

Several problems and defects may arise from the formation of Bénard cells— including flooding/floating and orange peel:

- Flooding/floating: In pigmented coatings, the pigment particles are caught in the turbulent flow and carried along. In situations where the pigments differ in mobility, floating may result. Such floating then becomes visible on the paint surface as color differences.
- Orange peel: Besides causing color changes, these eddy currents may also disturb the smooth surface of the coating; and a more or less pronounced structure (orange peel) is the result. Once again, silicones can often alleviate such problems, thereby enhancing performance.

13.8 Reactivity of Polysiloxanes (and Its Influence upon Intercoat Adhesion)

When polysiloxanes are synthesized in such a way that reactive groups are present within the molecular structure, the polysiloxanes can be incorporated into the binder. The reactive groups must first be suitable for entry into covalent bonding with the reactive groups of the resin system. The polysiloxane, based on its structure, must then be able to orient itself at the surface (Fig. 13.5). If the above two qualifications are fulfilled, a dimethylpolysiloxane interface is obtained which is not only firmly anchored in the surface layer, but also (depending on the concentration) no longer recoatable. The organic moieties with the reactive end-groups orient themselves in the binder and react there with the resin during the crosslinking process. An important factor is the appropriate choice of suitable reactive groups according to the resin type. In regard to certain two-pack systems, for example, specialized OH-groups and/or NCO-groups should be used. (For alkyd resins, carboxylic groups, and for systems curing by radiation, acrylic groups may be more appropriate.)

As shown below, a wide variety of beneficial properties can be obtained with tailored reactive polysiloxanes in UV-systems:

- No recoatability
- "Anti-graffiti effect" (especially useful for public utilities and buildings)
- Nearly permanent slip for furniture coatings and many other applications
- Improved soil release
- Excellent release properties for a wide variety of release papers

Crosslinkable Siloxanes-Mechanism

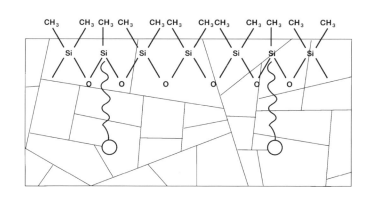

Figure 13.5 Covalent bonding as related to binder incorporation: Crosslinkable siloxanes

As a matter of comparison with previously mentioned polysiloxane derivatives—it is important to stress that *when products without the reactive groups are produced, then perfect recoatability can be obtained.*

13.9 Contact Angle and Surface Tension Parameters

13.9.1 Theoretical Aspects

If one regards a liquid as a plurality of molecules, then there are homogeneous attractive forces (intermolecular forces) acting upon the molecules within the liquid. As these forces operate simultaneously and homogeneously in all directions (at least in the "non-interface" zones), the forces are effectively balanced. However, at the interface of liquid to air, these forces are not equalized. In regard to the molecules at the surface, the primary forces acting are often those which are one-sided (directed to the internal liquid; sometimes the term "intra-liquid forces" is employed here) and which attract the molecules from the surface into the liquid. The liquid tends to reduce its surface; therefore, droplets and gas bubbles try to assume an ideal spherical shape. The surface tension is defined as force/unit of length and has the dimension mN/m. (Please note that 100 mN/m = 100 dynes/cm; several different surface tension values and dimensions are commonly employed in the industry.) The ring detachment method is the most common day-to-day method for determining surface tension values; a typical measuring apparatus is shown in Figure 13.6.

For measuring the surface tension, a platinum ring is first immersed in the test liquid and is then slowly drawn out at constant speed. The recorder then charts the resisting force, and the actual value of surface tension (maximum value) is empirically monitored (Fig. 13.7)

For determination of surface/interfacial tension by the ring pull-off method, the following standards—among others—may be employed: ASTM D 971, ASTM D 1331, DIN 53 914.

**Surface Tension
Measurement Apparatus**

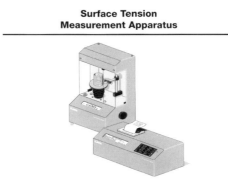

Figure 13.6 Platinum ring apparatus

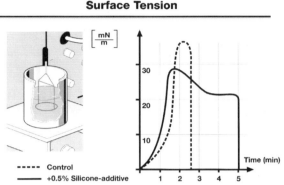

Laboratory Measurement of Comparative Surface Tension

Figure 13.7 Comparative surface tension

Comparative measurements are carried out by means of an inductive force transducer that converts the force into measurable electric signals which can be amplified, translated into germane values, and then displayed digitally. It is also a rather simple matter to connect a computer or a suitable recording apparatus to the instrument via built-in interface (RS 232). A lifting table raises and lowers the samples at a speed of 1.5 to 15 mm/min. In many aspects, surface tension and contact angle parameters are closely related. Although it is undoubtedly beyond the scope of this publication to define all the variables involved with contact angle, there are indeed a plethora of measurement methods. For reference purposes, three such methods are

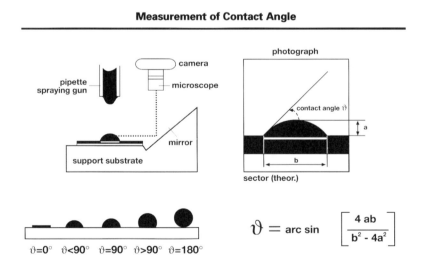

Measurement of Contact Angle

$$\vartheta = \text{arc sin}\left[\frac{4\ ab}{b^2 - 4a^2}\right]$$

Figure 13.8 Three laboratory methods

Table 13.4 Measurement of Angle on Paint Films

	Angle measured on paint film	Angle measured on paint film after cleaning with mineral spirits	△ Angle
Control	19.5	19.1	0.4
Product F	31.5	19.6	11.9
Product G	19.6	18.9	0.7

Test system: UV system
Measurement of angle: paint on paint

shown in an introductory fashion in Figure 13.8. In addition, a sampling of differential contact angle measurements, contingent on variable surface characteristics, is shown in Table 13.4.

13.9.2 Practical Aspects

Contact angle and surface tension measurements demonstrate that, by the judicious choice of polysiloxane, one can adjust interfacial tension properties in a substrate-specific manner. For instance, in the experiment shown in Figure 13.9, one zone of a paint surface containing an active silicone was left in an untreated condition. In contrast, another zone was wiped with a solvent, so that demonstrably different surface tension values were provoked (25 mN/m, unwiped area; 35 mN/m, wiped area). Of course, the reader may recall that an analogous experiment was described previously—in regard to ghosting. Upon these disparate zones, paint droplets corresponding in surface tension to that of the uncleaned clear coat were applied. If one compares the forms of the droplets in both areas, it is evident that the flattened—

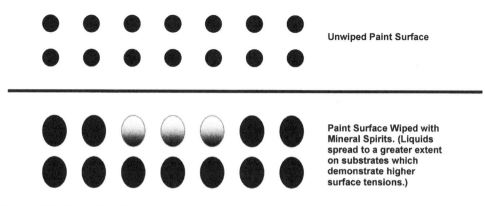

Figure 13.9 Spreading effect

and very dynamically spreading—droplets in the cleaned zone spread more. Indeed, the flattened droplets in the bottom half of the figure occupy significantly greater surface areas (200 + %) than do their counterparts (the ball-shaped or spherical droplets in the uncleaned zone) in the top half of the figure.

13.10 The Critical Surface Tensions of Substrates, Resins, and Solvents

Substrate wetting depends on the surface tension "interplay" between the substrate and the coating. The critical surface tension of a substrate is the surface tension which must be exhibited by the test coating to attain complete wetting of the substrate. Tables 13.5 to 13.7 display representative surface tensions of substrates, resins, and solvents. (Please note that even minuscule changes in surface composition or treatment may cause dramatic variations in surface tension; comprehensive ranges are not shown for all entries in the aforementioned tables.) Substrates with high surface tension are, of course, easy to wet, and those with low surface tension are more difficult to wet.

Table 13.5 Critical Surface Tension of Typical Substrates

Substrates	mN/m
• Glass	70.0
• Zinc-phosphatized steel	45.0–60.0
• Iron-phosphatized steel	43.0
• Tin plate	30.0–40.0
• Aluminium	33.0–35.0
• Steel (untreated)	29.0
• SMC Polyester	23.0

Table 13.6 Surface Tension of Resins

Resins	mN/m
• Melamine resin	57.6
• Epikote 828	46.0
• PVC	41.9
• Polybutylmethacrylate	41.0
• Alkyd (65% soya)	38.0
• Polyvinylacetate	36.5
• Polymethylmethacrylate	34.6
• Modaflow	32.0
• Dimethylpolysiloxane	19.8

Table 13.7 Surface Tension of Solvents

Resins	mN/m		
• Water	72.7		
• Butylcellosolve	31.5		
• Xylene (*o*, *m*, *p*)	30.0	28.6	28.3
• Ethylcellosolve	28.7		
• Toluene	28.4		
• Cellosolve acetate	28.2		
• Butyl acetate	25.2		
• *n*-Butanol	24.6		
• Mineral spirits	24.0		
• Methylisobutylketone	23.6		

In conjunction with resins, especially the more advanced varieties, there are vast differences in surface tension; this especially helps one understand some of the problems encountered when dealing with special substrates (such as plastics).

Perhaps one of the more interesting studies of critical surface tension involves the integration and graphical correlation of the following variables:

1. Critical surface tension
2. Flow properties
3. Surface tension
4. Degree of crosslinkage
5. Temperature
6. Time

In spite of the fact that entire textbooks could be written to explain and delineate the countless relationships existing among the above six variables, Figure 13.10 graphically depicts all of the major multivariate relationships involved. Of particular interest to the formulator is the achievement of the critical surface tension value, the absence of which can result in sub-optimal performance in regard to nearly all criteria.

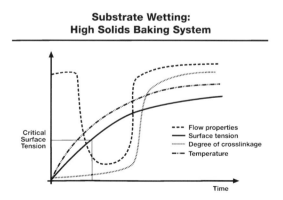

Substrate Wetting:
High Solids Baking System

Figure 13.10 Correlated relationships: Flow properties, surface tension, degree of crosslinkage, temperature, and time

Guide to Further Reading

Burrel, H., Ueber die Anwendung des Loeslichkeitsparameters in den Vereinigten Staaten, (22/5/62), Fatipec-Kongreß in Wiesbaden

Elser, W., Verlaufsmessung an Anstrichmitteln, Deutsche Farben-Zeitschrift, January (1962), Jahrg. 16

Fischer, E., Hamann, K., Elektronenmikroskopische Untersuchungen an Anstrichfilmen, *Farbe + Lack* (1957) 5

Funke, W., Vergleichende Untersuchungen über Trocknungsprüfmethoden für Anstriche, *Farbe + Lack* (1961) 12

Funke, W., *Schweizer Archiv für angewandte Wissenschaft und Technik* (1960) 8

Gettwert, G., *Lecture on Polymer Dispersion-based Silicate Paint*, Referat Ueber Dispersionssilikatfarben, Haus der Technik e.V., Essen November 28 (1989)

Hademar, H., *Messung des Deckvermoegens an Weißpigmenten, Fatipec-Kongreß* (1962), Verlag Chemie GmbH, Weinheim

Hummel, D.O., Scholl, F., *Infrared Analysis of Polymers, Resins and Additives: An Atlas* (1971) Halsted Press, New York

Kerner, H.T., *Foam Control Agents* (1976) Noyes Data Corp., Park Ridge, NJ

Preuss, H.P., *Paint Additives* (1971) Noyes Data Corp., Park Ridge, NJ

Technical Association of the Pulp and Paper Industry, *Paper Coating Additives* (1963) New York

Wapler, D., Über die Messung des Fließverhaltens von Anstrichmitteln, January/February (1958) Deutsche Farben-Zeitschrift

Zorll, U., Lichtmikroskopische Oberflächenuntersuchung von Lackschichten mittels einer Abdruckmethode, 8/196, Deutsche Farben-Zeitschrift

14 The Special Influence of Chemical Structure upon Performance: Advanced Principles of Molecular Design

14.1 Introduction

Since 1943, siloxane (or silicone) additives have played an integral role in coating performance; however, recent chemical innovations have greatly expanded this role, especially in the realm of environmentally friendly systems. A systematic analysis of molecular design is presented with special emphasis upon polyether-modified and polyester-modified polysiloxanes. The chemical determinants of silicone performance are outlined for environmentally friendly coatings, inks, adhesives and related applications. Not only will structure–performance correlations be discussed, but the following topics will be overviewed:

- System compatibility
- Surface tension effects
- Thermostability
- Wetting/levelling
- Slip/mar resistance

Silicon is one of the most frequently occurring elements in nature. As shown in an abbreviated fashion below, scientists tried very early to find parallels to the chemistry of carbon, with the intent of constructing a "quasi-organic" chemistry, so to speak. (Please note that exact course of "The History of Siloxanes" represents a sometimes rather highly debated field. Pioneering events, participating inventors, and crucial breakthroughs are often debated—with the end result sometimes being "a rotating reassignment" of certain scientific laurels. No attempt at comprehensiveness is made in the cursory listing below; the objective here is merely the provision of an exceedingly brief coverage of certain developmental events.)

The History of Siloxanes

"First" silanes (1872) A. Ladenburg
Siloxane Polymers (1940)   J. F. Hyde—Corning Glassworks USA
                           W.J. Patnode—General Electric USA
                           E.G. Rochow
                           B.N. Dolgov (U.S.S.R.)
                           K.A. Andrianow

Probable first large-scale commercial use in the paint industry (approximately 1950)

In 1872, Ladenburg produced some of the first elementary silanes and laid many of the foundations of "organosilane chemistry". Subsequent research received strong input from the intensive efforts of Kipping (University of Nottingham; approximately 1900–1940), who worked 45 years in the field of organosilicon and related chemistries, thereby supplying many of the early foundations for the development of the silicone industry.

Strange to say, but Kipping and many other researchers generally called the resinous oil- and gel-like compounds (which appear during synthesis of silanes) impurities. Early researchers were interested in obtaining pure, crystalline or distillable compounds. Accordingly, they tended to disregard the particular compositions which nowadays are of interest, without realizing their true significance.

Several years later, fundamental work was performed by Hyde at Corning Glassworks—and also by Patnode, Rochow, and numerous others at General Electric—in the field of the first polymeric compositions. At approximately the same time, similar research was performed in Russia by Dolgov and Andrianow. This work was inspired by the demands of the industry for heat-resistant plastics, especially in the electrical industry. The ensuing developments far surpassed the first goals. After the Second World War, a series of sudden breakthroughs launched the commercial utilization of polymer organosiloxane compounds for many different application fields. For instance, the use of silicones in the paint industry was described in 1952 by S.H. Bell as a solution for flooding/floating problems. In addition, many other researchers documented countless performance enhancing capabilities.

Figure 14.1 surveys selected application fields for polysiloxanes. Two of the smaller economic segments are represented by silicones for the paint industry and for the medical/pharmaceutical industries.

14.2 Synthesis of Siloxanes and their Intermediates

Figures 14.2 and 14.3 provide a brief review of the synthesis of silicones; in addition, Figures 14.4 and 14.5 display typical condensation and addition reactions.

The synthetic procedures and reactions shown thus far distinctly demonstrate that the basic chemistry is relatively complicated and can only be pursued by a few specialized firms. On the other hand, many companies of medium size have taken major steps towards further "segmented derivatization" of the basic raw materials (to supply smaller market niches with special products, thus to produce "tailor-made" silicones). Evolutionary advances in this relatively young industry are often quite rapid in nature, with the end result including not only market opportunities, but also the naturally occurring "growing pains" that accompany any industry in its infancy and youth. (One of the many challenges facing young disciplines—the siloxanes industry as well as the coatings industry—is illustrated, of course, by the recurrence of "competing definitions and/or nomenclatures" in various arenas. As alluded to before, the author has observed the occurrence of an absolute plethora of different spellings,

Application Fields of Silicones

Resins	• Paint • Electro* • Pharma • Building
Oils	• Paint / Electro / Pharma / Rubber • Cosmetics / Food • Plastics / Filler and Glass Fibers • Automobile / Textile / Paper / Foundry
Rubber	• Electro / Rubber / Building • Food • Automobile • Pharma / Medicine • Paper
Silanes	• Plastic • Rubber • Filler and Glass Fibers • Pharma
Defoamers	• Paint • Pharma • Textile • Metal

* Please note that a variety of industry-accepted "abbreviations" (such as "electro" and "pharma") have been employed in both the text and this chart.

Figure 14.1 Application fields

punctuations, and unit-to-unit conversion rules for certain technical terms and chemical nomenclature expressions in the industry. As paradoxical as it may seem, often even the accepted industry reference books and dictionaries differ greatly on certain conversion rules, spelling issues, "pluralization conventions", abbreviation shortcuts, hyphenation/punctuation rules and word-division protocols. Accordingly, several alternate conversion rules and/or grammatical usage practices—all equally correct—may indeed be employed by some readers. This commentary applies, of course, not only for this chapter—but also for all other chapters in this textbook.)

The polysiloxane functionalities used successfully in coating formulations obviously represent only a very small portion of the broad range of silicone chemistry. Silicones are widely used in other areas, e.g., plastic, rubber, pharmaceutical, paper, food, and cosmetic applications. Differences in molecular structure can be used to logically classify a multitude of products, and therefore help explain silicone behavior in coating systems.

14.3 Unmodified Polydimethylsiloxanes

The simplest products, as introduced in Chapter 11, are polydimethylsiloxanes, also known as "silicone oils" in some industry circles. Varying the chain length yields

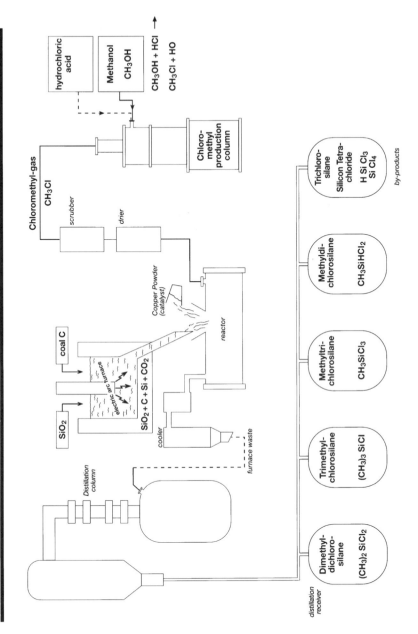

Figure 14.2 Synthesis (intermediate)

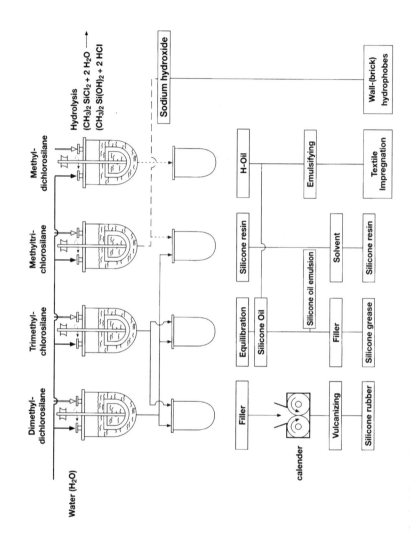

Figure 14.3 Synthesis (advanced)

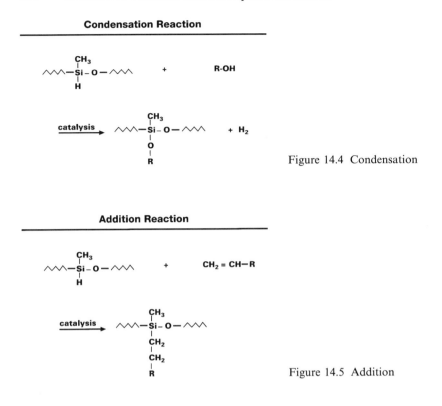

Figure 14.4 Condensation

Figure 14.5 Addition

silicones with different properties. With increasing chain length, these materials generally display higher viscosities. Therefore, viscosity is an excellent surrogate measure for the molecular weight of these simple products. Higher molecular weight usually means reduced solubility in coating systems and also less compatibility.

In addition to linear molecules, cyclic products are also possible. The dimethyl groups present in both types of silicones (linear and cyclic) provide the typical silicone effects of surface tension reduction and slip improvement.

Low molecular-weight polydimethylsiloxanes (number of "dimethyl units" $[n]$ < 60; please see Figure 11.5; of course, not all the definitional constraints of $[n]$ are presented here) and cyclic silicones are mainly used to control surface flow and to mask floating of pigments. Their viscosity is between 1 and 50 mPas. Such products may be volatile, so their usage levels and conditions are often critically sensitive. After evaporation from the paint film, they can possibly condense (e.g., inside the baking oven) and then contaminate other materials or substrates.

Products with higher molecular weight ($60 < n < 100$; viscosity, 50 to 100 mPas) can additionally be used as slip agents. Polydimethylsiloxanes with $n \sim 1200$ (viscosity $\sim 60,000$ mPas) often demonstrate just the right balance of compatibility/incompatibility with most common paint resins to be effective as defoamers without creating surface defects. These are the classical silicone defoamers.

Finally, those products with very high molecular weight (n > 1400; viscosity > 100,000 mPas) are so incompatible and insoluble that they consistently cause craters in nearly all paint systems; therefore, they are relegated to rather limited applications— and are generally used only to produce hammertone coatings.

14.4 Phenylpolysiloxanes

The pure methylphenylpolysiloxanes are of relatively minor importance in the paint industry, even though they are more paint compatible than pure dimethylpolysilox-anes. They influence levelling, but exert nearly no demonstrable effect on the other properties, such as flooding/floating, slip, and defoaming. (An example chemical structure is shown in Figure 14.6.) If only a few Si-atoms in the chain contain phenyl linkages, then products with graduated properties are obtained according to the proportion of dimethyl groups in the molecule. The end result is that slip, levelling, and defoaming can be properly controlled and influenced.

Phenylpolysiloxane

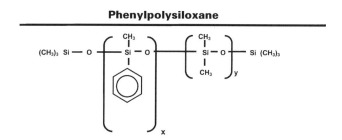

Figure 14.6 Phenyl moiety

14.5 Polymethylalkylsiloxanes

Polysiloxanes with methylalkyl groups (Fig. 14.7) instead of dimethyl groups still show typical silicone properties but in a less pronounced fashion. They reduce surface tension and improve surface slip, but not as strongly as the dimethyl products. Their degree of activity correlates with the chain length of the alkyl group (Fig. 14.8). Due to the higher surface tension of such products, coatings with these additives are generally easier to recoat in complex multi-layer systems. This type of silicone additive is oftentimes also found as the active substance in defoamers.

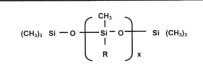

R =Alkyl (can be modified)   – (CH₂)ₙ CH₃

Figure 14.7 Methylalkyl groups instead of dimethyl groups

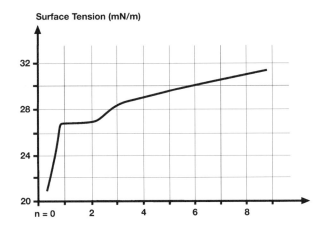

Figure 14.8 Activity correlated with alkyl chain lengths

14.6 Organically Modified Polysiloxanes

As explained above, the compatibility of polysiloxanes with coating systems can be controlled, in part, by varying the chain length. However, an infinitely more versatile way to achieve control is by introducing chemical modifications.

14.6.1 Polyethers

The most widely used modification is polyether chains (Fig. 14.9: please note the special EO/PO commentary in relation to the "R" group). Of course, the variation possibilities (Fig. 14.10) of the "R" group are of rather critical importance. As a matter of industry convention, certain "R" groups may sometimes be noted with subscripts, superscripts, and/or other accompanying "scripted symbols". To simplify

Polyether-modified Polysiloxane

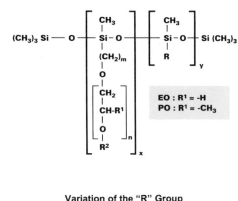

Figure 14.9 "R" groups in the polyether-modified polysiloxanes

Variation of the "R" Group

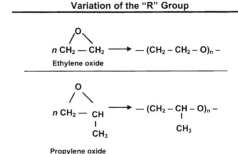

Figure 14.10 Chain variation

readability and ease of reference, though, this text will often refer to "R" groups without the use of such accompanying symbols.

This general chemical structure is so versatile that custom-made silicone additives with very specific properties can be synthesized. Dimethyl groups provide the low surface tension of silicones, so that by changing the ratio of dimethyl groups to organic modifications, the surface tension can be controlled. If, instead of dimethyl groups, methylalkyl groups are present, the resulting products will have considerably higher surface tensions. Thus, they will not be able to reduce the surface tension of the coating as strongly as the dimethyl products.

As mentioned before, the nature of the "R" group can be varied; accordingly, this moiety can consist of ethylene oxide (EO) units and/or propylene oxide (PO) units. (Obviously one could also expand this discussion—if there were no space limitations—to incorporate non-EO/PO entities.) Polyethylene oxide is strongly hydrophilic (polar), whereas polypropylene oxide is strongly hydrophobic (unpolar/non-polar); so it is obvious that the resultant ratio (EO/PO) can help control the overall polarity of the silicone additive. High EO content increases the polarity, so that the silicone becomes water-soluble and therefore more compatible in polar systems; however, the tendency to stabilize foam is also increased. High PO content, on the other hand, reduces water-solubility and increases defoaming properties. In addition to the EO/PO

The Variable of Chain Placement

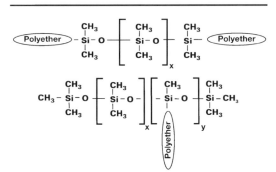

Figure 14.11 "End-chain" and "mid-chain" placement

ratio, the exact location of the polyether chains in the additive molecule is also critically important. As shown in Figure 14.11, innovative "end-chain" and "mid-chain" placement may be achieved in the more modern additive structures.

14.6.2 Polyesters

It is also possible to use polyester chains rather than polyether groups for modifying the silicone. The resultant polyester modifications (as shown in Figure 14.12) will also control the compatibility of the silicone product. One very important difference between a polyether-modified and a polyester-modified polysiloxane is thermal stability. Polyether-modified products are stable only up to the 150°C range (approximately 300°F).

Polyester-modified Polysiloxane

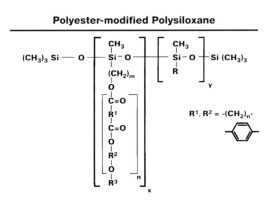

Figure 14.12 Polyester modification

| General structure | (CH$_3$)$_3$– Si — O —$\left[\begin{array}{c} CH_3 \\ | \\ Si-O \\ | \\ R_1 \end{array}\right]_x$ $\left[\begin{array}{c} CH_3 \\ | \\ Si- O \\ | \\ CH_3 \end{array}\right]_y$ Si (CH$_3$)$_3$ | | |

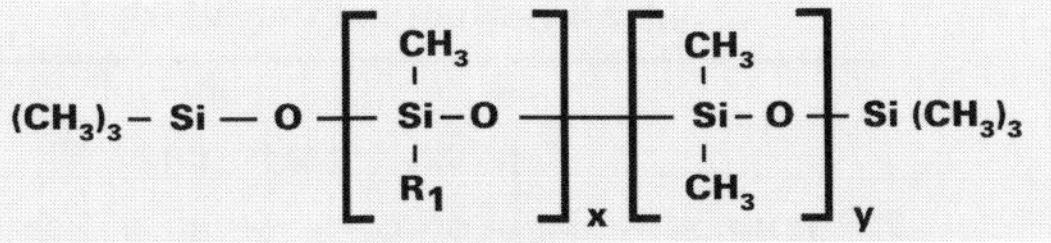

Properties \ R$_1$:	Type C -(CH$_2$-CH$_2$-O)$_n$-CH$_3$	Type A non-reactive -CH$_2$-(O-C-R$_2$-C-O-R$_3$)$_n$-O-CH$_3$	Type A reactive -CH$_2$-(O-C-R$_2$-C-O-R$_3$)$_n$-OH
Slip properties	1	2	2
Defoaming properties	4	3	3
"Antisilicone" properties	1	2	2
Levelling	1	1	1
Recoatability after overbaking	2	1	5
Intercoat adhesion <150°	1	1	5
Intercoat adhesion >150°	4	1	5
Reduction of surface tension	1	2	2

Estimation: 1 good / 5 poor

Figure 14.13 Properties of siloxanes

At higher temperatures, the polyether chain degrades by oxidation (which can be unequivocally proven by IR-spectra). The polyester modification, however, is stable up to the 220°C (430°F) range before degradation begins. A comparison of various polyester/polyether properties (including heat stability) is shown in Figure 14.13.

14.6.3 Aralkyls

Aromatic groups can also be used to modify the silicone backbone. The resulting products are, in general, aralkyl-modified silicones which display excellent thermostability. Aralkyl-modified polymethylalkylsiloxanes (Fig. 14.14) can combine excellent thermostability features with "higher surface tension effects"—meaning, in the special terminology of polysiloxane chemists, that surface tension is lowered to a value higher than that achieved with the dimethyl moiety. Of course, the higher surface tension effects are due to the presence of methylalkyl groups. If, in addition to the aralkyl and methylalkyl groups, dimethyl groups are also present, the surface tension will drop accordingly.

14.6.4 Reactive Siloxanes

Reactive products occupy a special position in the polysiloxane universe; a typical structure is shown in Figure 14.15. The general structure of the polysiloxane itself

Aralkyl-modified Polysiloxane

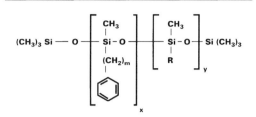

Figure 14.14 Aralkyl moiety

General Formula of Reactive Organopolysiloxanes

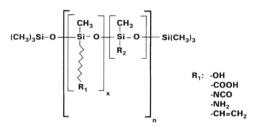

Figure 14.15 Reactive moieties

allows excellent orientation at the surface, while the resin-like polyester groups contribute to proper compatibility with the binder. The reactive groups are balanced and tailored to the curing mechanism of the paint binders. In summary, the above properties enable excellent cross-linking with paint resin itself. The results are unique surface properties which can generally be achieved only by silicone paints, i.e., permanent, wash-resistant slip, less dirt acceptance, and practically no recoatability.

14.7 Surface Tension Phenomena as Related to Siloxane Chemistry

Figures 14.16 and 14.17 display, in an introductory fashion, the development of (1) surface tension reduction in a typical UV coating, and (2) slip resistance reduction. (Please note that many different methods of discussing "slip phenomena" are utilized in the industry; no attempt at providing a tabulation of all "semantic variants" will be presented here.)

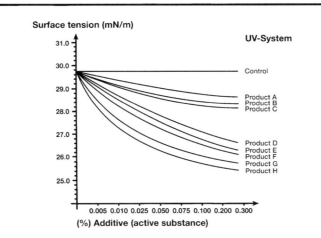

Figure 14.16 Surface tension

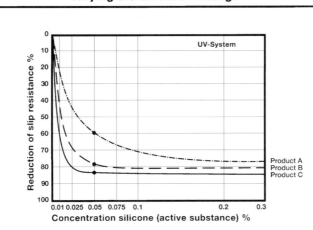

Figure 14.17 Slip

14.8 Special Aspects of Silicone Chemistry

14.8.1 Method of "Placement and Attachment"

In addition to the chemistry of the organic modification, the method of placement and attachment to the silicone backbone is also important. As mentioned previously, if the modification "R" is connected to the Si atom via an oxygen atom, the resulting silicone molecule is hydrolytically unstable. The Si–O–R bond is easily attacked under acidic conditions by water (even trace quantities are enough; see Figure 14.18). Such hydrolysis first splits off the organic modification from the silicone backbone, and thus the parent silicone loses its compatibility. As a result, turbidity may occur. In a subsequent step, the Si–OH groups can then condense, increasing the molecular weight of the silicone additive considerably. This will lead finally to a highly incompatible dimethylpolysiloxane which creates craters in the coating material. To prevent hydrolysis, it is necessary to use pendant Si–C (rather than Si–O–R) linkages.

In general, for most applications, silicones are expected to be non-reactive. If they turn out to be reactive, though, this may obviously cause problems with intercoat adhesion. Typically, non-reactive silicones can be removed, however, from the film simply by cleaning the surface with solvents (or perhaps with water if the system at hand is sufficiently water-soluble). Therefore, no permanent "silicone effects" (slip, scratch resistance) are achieved.

If "permanent effects" are, for whatever reason, deemed necessary and appropriate, then reactive silicones are required to cross-link with the binder and thus become an integral part of the resin system. Nevertheless, the silicone effects in this case are still not absolutely permanent because the silicone molecules are linked to the uppermost resin molecules in the coating. (Of course, whenever the terms "permanent", "absolute", and/or "irrevocable" are employed in this text, there may indeed be certain exceptions that can be identified—or even designed into—the system. In addition, whenever certain "system limits" are mentioned—such as discrete baking temperature limits, composition parameters, etc.—there may indeed be certain degrees of variability in selected systems.) As soon as these resin molecules are degraded

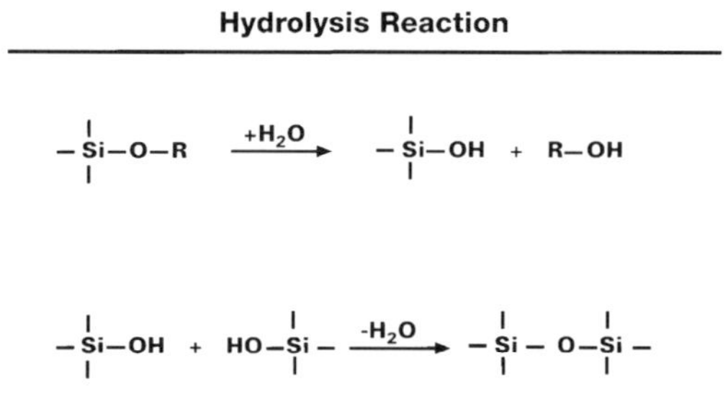

Hydrolysis Reaction

$$-Si-O-R \xrightarrow{+H_2O} -Si-OH \ + \ R-OH$$

$$-Si-OH \ + \ HO-Si- \xrightarrow{-H_2O} -Si-O-Si-$$

Figure 14.18 Deleterious hydrolysis

during the normal weathering process, then the silicone will also depart along with the resin. This factor must be considered for outdoor applications; however, it plays little or no role in regard to indoor applications.

14.8.2 Silicone Surfactants

Short-chain polydimethylsiloxanes (which contain only a few polyether chains as modifying groups) display the distinctive polar/non-polar structural dichotomy typical of surfactants; therefore, this explains why they are often called "silicone surfactants". (An example structure is shown in Figure 14.19: the reader may recall that brief mention of these products was also made previously.) Such products are particularly effective for reducing surface tension in water-based coatings, the surface-tension-reduction effect being much stronger than with conventional silicones exhibiting higher molecular weights. However, surface flow and surface slip are typically *not* influenced. (There is also no such influence in solvent-based systems.)

14.8.3 Stereochemistry

For organically modified polysiloxanes, the absolute number of organic modifications is only one of several important variables. The synthesis chemist must also carefully consider other factors—including the exact location and orientation of such modifications. For instance, as shown previously in Figure 14.11, at least two possible "molecular locations" exist:

- at the ends of the silicone backbone (linear)
- along the silicone backbone (comb or branched structure)

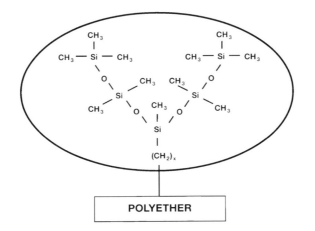

Figure 14.19 "Silicone surfactant"

The different structures above will, of course, lead to correspondingly different properties in the coating. The mobility of the silicones is especially influenced by steric factors. Although predictive indices of mobility are indeed available, one must—in practice—rely principally on empirical measurements.

Guide to Further Reading

Bell, S.H., *J.O.C.C.A.* (1952) p. 386

Boller, Dr. C., Wechselwirkung zwischen Furnier und Lack bei Andwendung von drei verschiedenen Lacksystemen, Gießen, *Farbe + Lack* (1966) 5

Carroll, B. (Ed.) *Physical Methods in Macromolecular Chemistry*, Vol. I (1969) Marcel Dekker, Inc., New York

Funke, W., Zorll, U., Elser, W., Über den Zusammenhang zwischen der Pigment-Bindemittel-Wechselwirkung und der Wasseraufnahme fester Anstrichfilme, *Farbe und Lack* (1966) 4

Ham, G.E., Vol. 1 (Vinyl) in the *Kinetics and Mechanisms of Polymerization Series*; Vol. 2, *Ring Opening*; Vol 3, *Step Growth*; Solomon, D.H., (1969) Marcel Dekker, Inc., New York

Haug, R., Elser, W., *Über die Beeinflussung der Wetterbeständigkeit von Holzanstrichen durch den Einlaßgrund*, 5/1966, Deutsche Farbenzeitschrift

Huggins, M.L., *Physical Chemistry of High Polymers* (1958) John Wiley and Sons, Inc., New York

Hunter, D.N., *Inorganic Polymers* (1964) John Wiley and Sons, Inc., New York

Lenz, R.W., *Organic Chemistry of Synthetic High Polymers* (1967) Interscience Publishers, Inc., New York

Kresse, P., *Farbe + Lack* (1966), p. 111

Margerison, D., East, G.C., *Introduction to Polymer Chemistry* (1967) Pergamon Press, Inc., New York

Mark, H., Tobolsky, A.V., *Physical Chemistry of High Polymeric Systems* (1950) Interscience Publishers, Inc., New York

Orr, E.W., *Silicones in the Coatings Industry: The Influence of Chemical Structure Upon Properties* (1995–1996) Lectures Available to FSCT Constituent Societies (special monograph series in support of FSCT lectures)

Saunders, K.J., *Organic Polymer Chemistry* (1973) Chapman and Hall, London

Segal, C.L. (Ed.) *High Temperature Polymers* (1967) Marcel Dekker Inc., New York

Tanford, C., *Physical Chemistry of Macromolecules* (1961) John Wiley and Sons, Inc., New York

Van Megen, H., Defoamers in Organic Solvent- and Waterborne-Paint Systems, *Faerg och lack Scandinavia* (1989) 7/8

Zorll, U., *Ein Gerät zur genauen Messung von Verlaufserscheinungen an flüssigen Anstrichschichten*, VII FATIPEC-Kongreß (1964) Verlag Chemie GmbH, Weinheim

15 The Reengineering of Interfacial Chemistry: An Integrated Approach to Acrylate and Silicone Chemistry

New acrylate and silicone chemistries have been developed to utilize analogous molecular design elements. In particular, alkyl side-chains and organic modifications (the latter employing enhanced polyether, polyester, and/or acrylate moieties) can be interjected into both acrylates and silicones.

Enhanced molecular design provides dramatically improved flow and levelling properties; accordingly, an in-depth analysis of flow, levelling, and interfacial tension is presented within the special context of comparative acrylate and silicone chemistry.

15.1 Acrylates versus Silicones: What's the Difference?

Acrylates and silicones—what are the similarities, and what are the differences? Is it possible to develop a synergistic approach to molecular design and performance improvement? The answers to these questions are provided within an integrated framework of new acrylate and silicone chemistries. Newly patented synergies have arisen; for instance, *special homopolymeric and copolymeric acrylate molecules (as shown in Figure 15.1) can be constructed in a tailored fashion, utilizing selected*

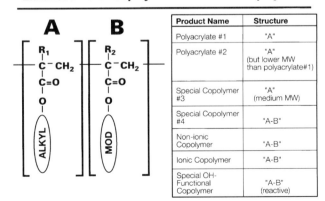

Figure 15.1 Molecular design synergies

molecular design elements which had previously been the rather exclusive domain of advanced silicone chemistry.

Specific molecular design elements—such as interface orientation, comparative chemical backbones, molecular weight, and polarity control—are discussed within both theoretical and practical contexts. The crucial differences between various acrylates are delineated; in addition, the physicochemical relationships of "flow" and "levelling" are defined and described. Finally, a concluding summary of selected performance enhancement principles is presented.

15.1.1 Comparative Mode of Action

How do acrylates and silicones perform? What are their differentiating characteristics? Why would one choose an acrylate over a silicone or vice versa? Although answers to the second and third questions will be provided in latter portions of this chapter, an introductory answer to the first question is shown in Figure 15.2. (Given the physicochemical complexities of acrylates and silicones, the multi-volume theoretical background principles of molecular design will obviously not be discussed in this overview chapter. Accordingly, only a cursory overview of selected theoretical concepts will be presented; the prime focal point will be that of providing a practical and useful framework for selecting acrylates and silicones to enhance performance.)

Within the comparative context of this chapter, the commonly used "flow and levelling" term is divided *into two distinct phenomena*. Is this really necessary from a physicochemical perspective though? Or is this merely a matter of semantics? There is a very logical basis for posing these questions; namely, many textbooks refer to "flow and levelling" as one and the same concept. Nevertheless, there does exist a very plausible argument, *at least in regard to understanding molecular design principles*, for systematically subdividing and defining the overall concept of "flow and levelling". Indeed, the physical mechanics of "flow" (within the special context of this chapter)

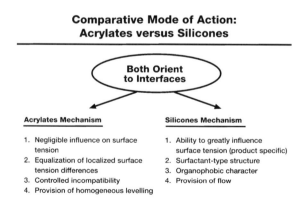

Figure 15.2 Comparative mode of action

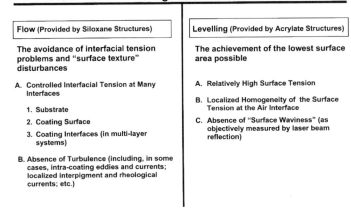

Flow Versus Levelling: Two Distinct Phenomena

Flow (Provided by Siloxane Structures)	Levelling (Provided by Acrylate Structures)
The avoidance of interfacial tension problems and "surface texture" disturbances	The achievement of the lowest surface area possible
A. Controlled Interfacial Tension at Many Interfaces	A. Relatively High Surface Tension
1. Substrate	B. Localized Homogeneity of the Surface Tension at the Air Interface
2. Coating Surface	C. Absence of "Surface Waviness" (as objectively measured by laser beam reflection)
3. Coating Interfaces (in multi-layer systems)	
B. Absence of Turbulence (including, in some cases, intra-coating eddies and currents; localized interpigment and rheological currents; etc.)	

Figure 15.3 Differentiating characteristics

can be understood as phenomena which are distinct and, at times, totally different from their "levelling" analogs. In elucidation of this point, the differentiating characteristics of flow *versus* levelling are outlined in Figure 15.3.

15.1.2 Flow Versus Levelling: Two Distinct Phenomena

Because "flow and levelling" are such distinct and separate phenomena—different base chemistries were historically developed in the traditional realms of silicone (flow) and acrylate (levelling) technology. Utilizing the two resultant chemistries, two divergent classes of products were first introduced; however, modern synthetic procedures allow increased integration and control of molecular design, regardless of which class of products is being synthesized. To begin with, analogous organic modifications—such as tailored polyether, polyester, and aralkyl moieties—can be interjected into *both* acrylates and silicones. Of course, such modifications represent only a meager first step; additional improvements include analogous block copolymer structural elements, of which introductory examples have already been displayed. (Comparative chemical structures will be discussed at length in subsequent sections of this chapter.) Further improvements include tandem polarity adjustment procedures and alkyl modifications.

The chemical backbone structures of both acrylates and silicones, as displayed in Figure 15.4, contain many similarly situated backbone and pendant structures; so it is certainly not surprising that analogous synthetic procedures can often be employed. Section 15.2 overviews selected molecular design elements, with special emphasis placed on the practical implications of the resultant molecules. An integrated analysis of flow and levelling principles—along with a brief compendium of performance enhancement principles—will be presented.

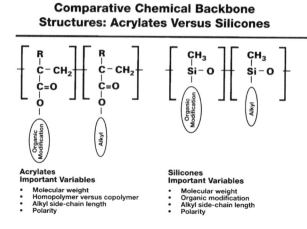

Figure 15.4 Comparative chemical backbone structures

15.2 Enhanced Molecular Design: How and Why?

Both silicones and acrylates orient to the interface; however, silicones, in contrast to acrylates, appreciably reduce surface tension. The distinctly alternating Si–0 backbone (in contrast to the traditional carbon-based backbone of acrylates) is obviously a major contributor to the silicone's organophobic, surfactant-like character. Surface tension can be lowered, and surface slip can be increased. (Of course, one can also quite easily design silicones to lower surface tension *without* influencing slip; nevertheless, no attempt at an encyclopedic enumeration of all synthetic possibilities will be presented within the scope of this overview publication. More detailed information about silicones can be found in Chapters 11 to 14. The objective of Chapter 15, however, is the comparison, juxtaposition, and final integration of silicone and acrylate technology.) *Acrylates, in comparison to silicones, are quite effective in equalizing localized surface tension differences, yet they are not specifically designed to lower surface tension.* Utilizing the initial example of a modified polysiloxane, a discussion of selected molecular design elements—some facets of which have been presented in previous chapters, but in a rather different context—will now be presented as a springboard to a broader discussion. Afterwards, both acrylates and silicones will be presented, with particular focus on comparative polarity adjustment procedures and side-chain modifications.

15.2.1 Silicones: Building Blocks

As shown in Figure 15.5, the "x" building block of a typical polysiloxane can include modifications with polyether, polyester, and related side-chains; nevertheless, it cannot

General structure

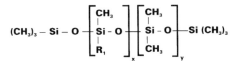

Figure 15.5 "x" and "y" building blocks

be emphasized enough that one must always utilize *pendant carbon groups as the linkage* between the polyether side-chains and the alternating silicon/oxygen backbone structure. (Please note that several different nomenclature variations, especially in regard to the building blocks, may be encountered in the industry.) Obviously, pendant oxygen group linkages attached directly to the backbone are relatively unstable and should always be avoided.

The relationship or proportion of dimethyl groups to organic modifications (y to x) also allows one to control the degree of compatibility. Simultaneously, this relationship exerts a direct influence upon surface tension (the greater the number of dimethyl groups available, the lower the surface tension—as a rule of thumb). In addition, one can also replace (either partly or entirely) the methyl groups with long chain alkyl groups, thus moderating the silicone's ability to lower surface tension. Accordingly, where the variable "y" (as applied to silicones) is shown in Figure 15.5, one could employ both methyl and/or alkyl groups as molecular sub-components.

Another important route of performance modification includes the controlled variation of the polyether chains themselves. An essential variable here is the polarity of the structural elements. The pendant chain—at least in simpler scenarios—may contain ethylene oxide (EO) and/or propylene oxide (PO) units. As mentioned in a previous chapter, polyethylene oxide is very hydrophilic (polar); however, polypropylene oxide is comparatively hydrophobic (non-polar). Through utilizing various proportions of EO to PO, one can control or modify the polarity of the entire silicone additive. For example, a higher proportion of EO raises the polarity so that the additive is water soluble and thus more compatible in very polar paint systems. Nevertheless, the tendency toward foam stabilization in the aforementioned case increases. On the other hand, a higher proportion of propylene oxide (PO) reduces both the water solubility and the foaming tendency.

15.2.2 Acrylates: Building Blocks

Thus far, the above discussion has merely set the stage for a broader presentation of the similarities between synthesizing acrylates and silicones. Some of the basic building blocks—ranging from elementary pendant groups (polyethers and polyesters) to alkyl group modifications—have been introduced. Acrylate molecules (as shown in Figure 15.6) can often take advantage of the best that silicone chemistry has to offer.

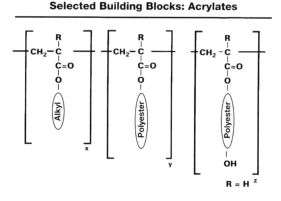

Figure 15.6 Acrylates

Molecular weight, polarity, and alkyl chain length can be tailored for homo-polymeric and copolymeric acrylates. In the *homopolymeric* polyacrylate family, important products include middle- or high molecular-weight poly-*n*-butyl acrylate derivatives. In regard to *copolymers*, levelling agents such as the copolymerization products of 2-ethylhexyl acrylate with various comonomers are of importance. Of course, many other chemical building blocks and components, in addition to the aforementioned, are available.

Can such structures be modified in the same manner as silicones? Yes, to a large extent, they can. As shown in Figure 15.7, many completely analogous structural elements can be interjected into acrylate molecules. Corresponding modification scenarios exist for both acrylates and silicones; however, there sometimes exist specialized polarity adjustment methodologies (certain salt structures, for instance) in advanced acrylate chemistry. Of course, organic modification is by no means a unidimensional variable; more than mere polarity control and related attributes can be adjusted by judicious design of the organic modification component. (Additional information will be provided in subsequent discussions of structure–performance enhancement.) The acrylate/silicone analogy can be extended not only to polarity adjustment procedures, but also to alkyl side-chain modifications. Because of a wide variety of factors—one of which being the differential molecular content of acrylates versus silicones (silicon versus carbon atoms)—shorter chain alkyl groups *in silicones* provide improvement in flow, slip, and substrate wetting; however, a much different effect (levelling) occurs in analogous acrylate structures.

In regard to longer chain alkyl groups, silicones can exhibit defoaming action— *along with minor influences on levelling behavior*; concomitantly, there is not as much reduction in surface tension as with the shorter alkyl moieties. (So, this obviously proves that flow and levelling effects are not always mutually exclusive of one another in the silicones family of additives; this is one instance where one can actually observe some measure—however minor—of *levelling* improvement in a silicone molecule.) In contrast, many acrylates often exhibit *considerably more pronounced levelling features*

Comparative Polarity Adjustment Procedures

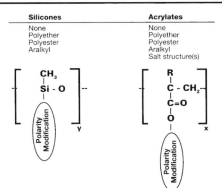

Silicones	Acrylates
None	None
Polyether	Polyether
Polyester	Polyester
Aralkyl	Aralkyl
	Salt structure(s)

Figure 15.7 Analogous structural elements

when shorter, rather than longer, alkyl chains are present. Acrylates demonstrate improved "reflow" (including, for instance, the possibility of improved "anti-popping" properties) when longer alkyl chains are present. Other important determinants of levelling behavior undoubtedly exist, so alkyl chain length is only one variable of many, yet its importance should not be underestimated.

In summary, an exceedingly broad palette of performance parameters can be enhanced (as shown in the comparative structure–performance summary in Figure 15.8) by a concerted approach to alkyl chain length in both acrylates and silicones.

Comparative Alkyl Side-Chain Lengths:
Structure-Performance Relationships

Silicone Backbone		Acrylate Backbone	
Shorter Alkyl	*Longer Alkyl*	*Shorter Alkyl*	*Longer Alkyl*
Flow	Defoaming	Levelling	"Anti-popping"
Slip	"Levelling"		(good "re-flow")
Substrate wetting	Less surface tension reduction		

Figure 15.8 Comparative structure–performance relationships

15.2.3 The Twin Variables of Molecular Weight and Polarity Control

Thus far, only *single* variables—ranging from the basic building blocks of the backbone to polarity adjustment procedures and alkyl side-chains—have been discussed; at this point, the most fruitful approach to continued analysis lies in the delineation of *multi*variable factors and effects (multivariate analysis). Accordingly, a cursory overview of the combined utility of two randomly chosen variables (molecular weight and overall polarity control in acrylate molecules) will be briefly examined.

Beyond a shadow of doubt, a multivolume treatise would not suffice to cover every aspect of multivariate analysis—especially with all of its attendant statistical protocols—so rather than concentrating on the comparatively "dry" aspects of statistical analysis, *only the practical end results will be presented.* Furthermore, for simplicity's sake, the overview analysis presented here will concentrate exclusively on acrylate analogs.

As shown in the introductory diagram in Figure 15.9, a wide spectrum of molecular-weight (represented by the "*y*" axis) and polarity-control (represented by the "*x*" axis) permutations can exist; not only are there homopolymers and copolymers, but there are also ionic and non-ionic species. Certain polymers are designed for solvent-based systems, whereas others are more suitable for aqueous systems. Within the bi-functional framework of molecular weight and polarity, selected "grid zones" can furthermore be associated with various formulating and/or performance criteria.

Of course, the graphical depiction of two-variable analysis is only a first step; all of the critically important aspects of practical formulation chemistry are missing so far. For instance, *exactly which performance criteria (such as levelling, anti-cratering, air release, anti-popping, etc.) are represented by the respective areas of the grid?* Which products are designed for special application areas? In response to these questions, a

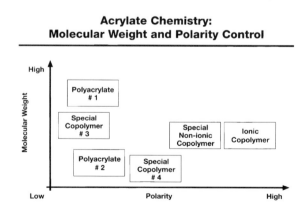

Acrylate Chemistry:
Molecular Weight and Polarity Control

Figure 15.9 Introductory molecular weight and polarity control variables

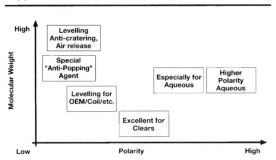

Figure 15.10 Practical applications grid

specially prepared grid, entitled "Practical Applications for Enhanced Acrylate Performance," is shown in Figure 15.10.

Many additional variables and "grid axes" could be incorporated into the grid paradigm. For example, one could add respective axes and/or *sub*-axes for alkyl length, polarity *type* (polyethers versus polyesters, aralkyls, and salt structures), and even copolymer orientation and reactivity. All such variables are of critical importance, and thus allow the formulator the ability to utilize—in acrylate moieties—many molecular design elements which had previously been the rather exclusive domain of advanced silicone chemistry.

15.3 An Integrated Analysis of Flow and Levelling

Given the performance utility of tailored acrylate and silicone structures, an integrated analysis of "flow" and "levelling" is necessary to properly guide and fine-tune the choice of an appropriate additive. Once again, the scope of this overview publication prohibits encyclopedic coverage of all physicochemical aspects, yet additional basic tenets are reviewed and fine-tuned—within the special framework of this chapter's perspective—in this section. First, flow principles (within the context of polysiloxane technology) will be covered, followed by a comparative synopsis of levelling principles (within the context of acrylate technology).

15.3.1 Flow—Part 1: Eliminating Intrafilm Turbulence

The solvents evaporating from paint carry a low viscosity medium of lower density from the deeper zone to the surface. This medium spreads out, the solvents evaporate, and the system becomes more viscous; furthermore, this medium acquires a higher density and sinks back into the bulk of the paint, which is richer in solvent. The pigments are carried along and whirl about in continuous eddies. As the paint dries and becomes more viscous, the movement gradually comes to a halt. The resultant "cell structures" on horizontal surfaces are obviously denoted as the rather ubiquitous *Bénard cells* that have been previously introduced. On vertical surfaces, the further deformation of these special cells produces *silking*.

Silicone additives enhance performance by orienting to the surface (in this case, the air/liquid interface). They form a micro-layer which immobilizes the system at the surface, so there are no zones of differential surface tension and, as a result, no Bénard cells or defects caused by air-draft sensitivity.

15.3.2 Flow—Part 2: Eliminating Intrafilm Surface Tension Differences

A liquid tends to wet areas of higher surface tension. With this in mind, one can now explain how craters may form during a spray operation. If the spray dust has a lower surface tension than the already applied liquid paint, then craters occur. This is because the spreading direction is *from* areas of lower surface tension *to* areas of higher surface tension. By pushing forward the already applied coating material, a crater is formed.

On the other hand, if the overspray material displays higher surface tension—then such material will not cause craters. It will remain on the surface of the already applied coating. (This abbreviated discussion does not include possible intervening variables—such as gravity, chemical interaction, etc.)

15.3.3 Flow—Part 3: Interfilm Effects

Next, reactive versus non-reactive modifications are compared; cross-linking the siloxane with the paint resin prevents recoatability. In contrast, non-reactive modifications will not cross-link with the resin; therefore, they will not anchor themselves to the paint resin in the first layer. Their resultant freedom to migrate into the second coat allows recoatability. Through utilizing knowledge of exact chemical structure (as described in previous chapters), complete control of intercoat adhesion can be achieved.

15.3.4 Flow—Part 4: Summary of the Chemistry and Physics of Enhanced Flow Structures

Alkyl groups, organic modifications, and tailored raw materials allow drastically increased control of *flow* properties. As discussed in the next section, many of the aforementioned principles (at least in the molecular design arena) are quite versatile, and can therefore also be employed in the synthesis and design of *levelling* agents.

15.3.5 Levelling—Part 1: Two Prerequisites for Proper Levelling

There are two primary prerequisites for optimized levelling: First, a *high* surface tension (which, in order to assure proper substrate wetting, is nevertheless lower, in most cases, than the surface tension of the substrate); second, a *localized homogeneity* of the surface tension at the air interface. Many coatings without levelling additives display not only improper levelling (the well-known "orange-peel effect"), but also crater-susceptibility. Such phenomena can be explained by the film disturbances brought about by surface tension gradients.

15.3.6 Levelling—Part 2: "Orange Peel", Craters, and the Laws of Physics

The laws of physics dictate that coating moieties on the surface *must migrate toward areas of higher surface tension*. Such migration can sometimes result in a highly structured, wavy surface (orange-peel effect). In addition, craters may form when the localized surface tension differences are great enough for the low surface tension areas of the paint film to actually break open. (Several other physical phenomena may also contribute to crater formation; some are even integrally related to both flow and levelling behavior.) Dynamically forming craters—with diameters sometimes growing from 0.1 to 20.0 + mm—can often be formed in such a manner that they penetrate to the substrate. At times, these surface defects are practically independent of the substrate choice; accordingly, they cannot always be classified as typical substrate wetting disturbances. In contrast, they are often the result of extreme *levelling* disturbances.

15.3.7 Levelling—Part 3: How Levelling Additives Function: Mode of Action

Modern-day levelling additives function through their influence upon interfacial properties. They display negligible surface tension influences; nevertheless, they do equalize localized surface tension differences, thus creating the ideal prerequisites for even and homogeneous levelling.

In addition, there also exist special categories of levelling disturbances which are caused by localized *rheological* differences (especially in regard to flocculated pigment agglomerations/accumulations). Even though the aforementioned pigment aberrations are certainly not the main focal point of this publication, properly employed acrylate additives help alleviate many levelling problems associated with flocculation and non-optimal pigment behavior.

The method of action often demonstrated by levelling additives is based upon a certain controlled incompatibility in the paint film; such beneficial and controlled incompatibility allows the additives to concentrate at the interfaces, thereby beneficially influencing interfacial tension. As a result, enriched additive concentrations form at the interface, and *localized surface tension differences are equalized.*

15.3.8 Levelling—Part 4: Summary of Problems Solved by Specially Modified Acrylates

Accordingly, a wide variety of levelling products can be developed, encompassing homopolymers and copolymers—ionic or non-ionic in nature. When desired, special reactivities can also be introduced. Performance enhancement, as shown in Figure 15.11, can be dramatically improved in both solvent-based and aqueous systems.

The most optimal methodology by which acrylates can enhance performance involves the design principles discussed herein. Enhanced homopolymeric and copolymeric acrylates can utilize the following molecular modification methodologies from the polysiloxane chemistry realm:

- Tailored organic modification
- Enhanced polarity adjustment
- Controlled and balanced molecular weight
- Specially designed alkyl chain length

As discussed previously in this chapter, several permutations—especially those involving multivariable scenarios—can greatly contribute to the possibility of utilizing the best of both acrylate (as shown in the left half of Figure 15.12) and polysiloxane (as shown in the right half of Figure 15.12) chemistries. *The end result is enhanced control of performance in both of the original chemistries.*

Problems Solved by Specially Modified Acrylates

Solvent-Based Systems	Aqueous Systems
• low surface tension	• high surface tension
• uneven surface	• substrate wetting
• craters	• craters
• levelling at low film thicknesses	• boiling marks
	• foam
Comment	*Comment*
Acrylates provide "fine-tuning."	*Acrylates are often absolutely essential.*

Figure 15.11 Acrylates in solvent-based and aqueous systems

Integrated Molecular Design Components

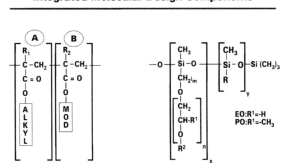

Figure 15.12 Integrated design

15.4 Performance Enhancement Principles

Although "flow" and "levelling" are indeed distinctly unique phenomena, a fully integrated approach to molecular design often provides the most optimal results. This certainly does not necessarily imply that silicon atoms appear in the backbone structures of acrylates, or that pure acrylate linkages appear within polysiloxanes. However, there are several analogous modification methodologies (as shown in the synopsis presented in Figure 15.13) in which acrylate and silicone chemistries share common ground. New and exciting synergies have arisen. Through judicious application of molecular design principles, performance can be dramatically enhanced. For instance, special homopolymeric and copolymeric acrylate molecules can be constructed in a tailored fashion, utilizing selected molecular design elements which had previously been the rather exclusive domain of advanced silicone chemistry.

ACRYLATES AND SILICONES:
Synergistic Molecular Design Aspects

---

1. Organic Modification Possibilities (Generally organic in nature)

a. Silicones	b. Acrylates
(i.) None	(i.) None
(ii.) Polyether	(ii.) Polyether
(iii.) Polyester	(iii.) Polyester
(iv.) Aralkyl	(iv.) Aralkyl
	(v.) Salt structures

---

2. Polarity Adjustment

Commentary:
In regard to both silicones and acrylates, the organic modification possibilities delineated in Section #1 above can serve as a prime method of polarity adjustment.

---

3. Molecular Weight

Commentary:
Generally speaking, molecular weight must be considered within the overall context of molecular design. Of course, one rather crude "rule of thumb" states that higher molecular weights are often correlated to increased levels of incompatibility; however, a wide variety of intervening variables (such as those enumerated in Sections 1, 2 and 4 of this chart) may ultimately exert far greater influences on final performance.

---

4. Comparative Alkyl Chain Lengths

a. Silicones	b. Acrylates
(i.) Shorter (increased flow)	(i.) Shorter (increased levelling)
(ii.) Longer (increased defoaming action/minor influence on levelling/not as much reduction in surface tension)	(ii.) Longer ("anti-popping" effects/good "re-flow")

Figure 15.13 Synergistic aspects

Guide to Further Reading

Allyn, G., *Acrylic Resins, Unit 17 of the Federation Series on Coatings Technology* (1971) Federation of Societies for Coatings Technology, Philadelphia, PA

Funke, W., Die Prüfung von Anstrichen unter variierten Klimabedingungen (Mehrpunkt-Methodik), Jahrg, *Farbe + Lack* (1963) 2, p. 69

Horn, M.B., *Acrylic Resins* (1960) Reinhold Publishing Co., New York

Martens, C.R., *Alkyd Resins* (1961) Reinhold Publishing Co., New York

McGregor, R.R., *Silicones and Their Uses* (1954) McGraw-Hill Book Co., New York

Meals, R.N., Lewis, F.M., *Silicones*, (1959) Reinhold Publishing Co., New York

Merz, O., *Gibt es einen wahren Fortschritt in der Lackprüftechnik?* January (1957) Deutsche Farben-Zeitschrift

Ranney, M.W. (Ed.) *Silicones, Vol. 2, Coatings, Printing Inks, Cellular Plastics, Textiles and Consumer Products* (1977) Noyes Data Corp., Park Ridge, NJ

Smith, D.A. (Ed.) *Addition Polymers: Formation and Characterization* (1965) Plenum Press, New York

Part IV

Defoaming
(The Removal of Unwanted Interfaces)

16 An Introduction to Foam Phenomena

16.1 Background

Foam can be characterized as an extremely large interface of gas to liquid. Furthermore, foam always involves the closed incorporation or "encapsulation" of a gas (generally air) within the aforementioned interface. In regard to environmentally friendly coating systems, foam can pose special problems. Oftentimes, the interfacial tension parameters of environmentally friendly coatings present myriads of challenges which may seldom occur in conventional coatings systems.

As a result, new defoamer technologies have been systematically developed to alleviate the special foam problems occurring in the most modern systems. An introduction to defoamers and air release agents is presented within the context of advanced chemical structures and new performance mechanisms.

Recent chemical advances allow increased foam control; an analysis of both theory and practical application data is therefore presented. First, the bifunctional physical and chemical determinants of foam are described. Next, defoaming mechanisms are examined for typical solvent-based, water-based, and latex systems. Newly patented technology is described along with a step-by-step analysis of balance and compatibility. Not only will polysiloxanes (and their organic modifications) be discussed, but novel polymer derivatives will also be introduced.

16.2 What is Foam?

Foam, as briefly mentioned before, is defined as a fine distribution of a gas, normally air, in the liquid phase. A characteristic feature of foam (as compared to other physical states) is the extremely large gas/liquid, lamella-type interface separating the gas bubbles from one another. Because of the basic physical and thermodynamic laws involved, every liquid attempts to achieve the lowest possible energy state. Foam, though, is a prime example of a temporarily displaced "energy anomaly", so to speak. As such, foam correlates with an elevated energy state, so one can conclude that foam is possible only when foam stabilizing factors are present. Furthermore, as a logical corollary, one can quite readily deduce that the prevention and elimination of foam must be integrally linked to lower energy states.

Regardless of the outcome of one's energy state analysis, one thing is for sure: foam is not normally a desired property in coatings. Deleterious foam can obviously occur, for instance, during the production stage, leading to inaccurate loading of the production vessel. During draw-off, it may prevent correct filling. At the application stage, foam can cause surface defects leading to weak points in the film, thus impairing final coating performance. Therefore, a defoamer is an intrinsically important component of most coating formulations.

16.3 The Mechanism of Foam Formation

One of the most common prerequisites of foam production is the presence of surface active chemicals whose molecules are composed of both hydrophilic and hydrophobic structural units. Such bi-functional chemicals orient strongly to the liquid/gas interface (Fig. 16.1), and reduce the surface tension.

As shown in Figure 16.2, foam can arise from a variety of different sources; it functions not only as an optical disturbance, but also as a hindrance to the proper development of the coating's decorative and protective roles. Accordingly, with the

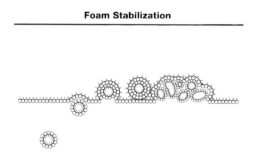

Figure 16.1 Foam: Schematic depiction

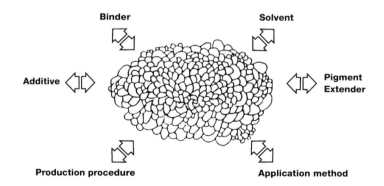

Figure 16.2 The origins of foam

intent of learning enough about foam formation to prevent and curtail it, selected physicochemical principles and observations will now be presented.

By carefully observing foam throughout its lifetime, it is possible to tabulate gradual changes in its structure. Shortly after formation, the foam has a relatively high liquid content, and this "wet" or "spherical" foam contains numerous globular bubbles which exert little or no influence on one another. Nevertheless, as described in the next section, this rather transitory foam is rapidly evolving.

16.4 Drainage, Charge Repulsion, and Elasticity

The liquid begins to drain from the foam (drainage effect) and the lamella becomes thinner as the bubbles are drawn closer together, gradually pressing against one another and forming polyhedral shapes. The resultant structures are denoted as "*dry*" or "*polyhedral foam*". The aforementioned liquid drainage has the effect of destabilizing the foam, because such drainage leads to thinner and thinner foam interfaces which would simply rupture if there were no counteracting physicochemical effects.

One counteracting effect can be traced to the chemical structure of the surfactants. In aqueous systems, the hydrophilic groups on the surfactants are of an ionic nature. The two interfaces of the lamella, which are covered with surfactant, come closer and closer together because of the "*drainage effect*". As shown in Figure 16.3, the equally charged ionic groups of the surfactant repel one other, thus preventing not only further drainage, but also the rupture of the lamella.

A further stabilizing effect is provided by the elasticity of the foam lamella. If the lamella is stretched, then eventually—at a certain critical length—a reduction in surfactant concentration at the interface can be observed. This is because the number of surfactant molecules in a given volume of liquid is finite and because the larger surface area must therefore exhibit a lower concentration. This results in an increase in surface tension, and the film tries to shrink like a rubber skin, thus restoring the surfactant concentration to a foam stabilizing level. Such dynamic shrinkage of thin

Foam Stabilization by Electrostatic Repulsion

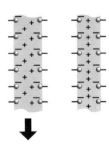

Figure 16.3 Stabilization

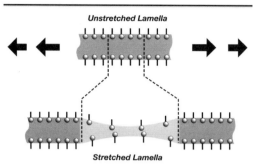

Figure 16.4 Lamella phenomena

liquid films is denoted as, in some circles of the industry, "Gibbs elasticity" and is shown in Figure 16.4.

Although there are certainly other effects which could theoretically influence foam stability, it can generally be concluded that *liquids foam only in the presence of surface active chemicals which stabilize the foam by electrostatic repulsion, Gibbs elasticity, and/ or other intervening physicochemical phenomena. (Dozens of such phenomena exist; several hundred pages would be required to describe all of them.)*

Almost without exception, most paint system components can affect foam behavior either positively or negatively. In addition, the substrate and the application methodology can also affect foam behavior. Of course, foam formation, avoidance, and/or removal are highly "situation specific". For instance, a particular spray application may result in excellent film properties; nevertheless, the utilization of the exact same paint system in a curtain coater operation may allow foam problems to occur.

Foam bubbles generally rise through the liquid to reach the air/liquid surface. According to one permutation of Stokes' Law, the rate of rise is dependent upon the radius of the bubbles and upon the viscosity of the liquid. (Larger bubbles obviously rise faster; more about the exact rate of rise will be discussed in the next chapter.) When a typical gas bubble reaches the surface, liquid then flows down and consequently out of the lamella. This flow process reduces lamella wall thickness, begins the dynamic process of forming geometrically stable hexagonal structures (as shown in Figure 16.5), and is denoted as the "drainage effect". (Please note that stable hexagonal structures must be formed from rather unstable polygons which originally have fewer than six sides; accordingly, Figure 16.5 displays a pentagon being "molded and melded", so to speak, into the final stages of stable hexagon formation.) Beneath 10 nm wall thickness, the lamella loses its integrity and the foam bubble breaks.

Of course, if coating behavior were to precisely follow the aforementioned energy reduction paradigm, then there would be absolutely no foam problems, since stable foam could not form. This is, for example, generally the case with pure liquids; pure liquids simply do not foam—at least not for more than a fraction of a second. In order

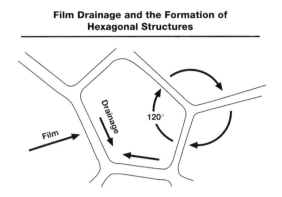

Figure 16.5 Flow

for foam bubbles to be capable of forming, then foam-stabilizing substances—as mentioned before—must be present in the liquid phase. Every paint formulation (aqueous, solvent-free, or solvent-based) contains a multitude of foam-stabilizing substances of varying origin and chemical structure. As a result, all formulations are capable of stabilizing foam.

16.5 The Requirements of an Effective Defoamer

As shown in Figure 16.6, an effective defoamer must overcome *both* Gibbs elasticity and electrostatic repulsion. Since both of the aforementioned foam stabilizing factors involve the maintenance of a relatively high energy state, then they serve as foam

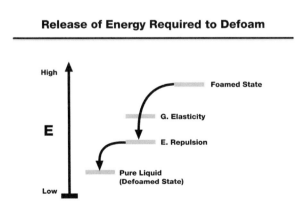

Figure 16.6 Energy diagram

release barriers or thresholds. Proper defoaming occurs only when one permeates the barriers, releases energy, and moves lower in the energy diagram shown in Figure 16.6. There are naturally countless other foam-related phenomena; nevertheless, additional theoretical information will not be discussed in this chapter.

16.6 "Defoamers" versus "Air Release Agents"?

Often one refers to "defoaming" whenever the removal of both encapsulated and/or "surface" bubbles is required; however, semantic distinctions among various bubble removal processes are sometimes made. To begin with, the gas bubbles must somehow reach the surface. This very process of reaching the surface is referred to as "air release"; on the other hand, the destruction of air bubbles already *on the surface* represents the actual "defoaming". In practice, though, this strict nomenclature distinction is not always observed. One should also consider that it is not always possible to unequivocally define the working mechanism of additives as either "defoaming" or "air releasing" in nature. As a result, subsequent discussions of foam will generally employ the term "defoamer", even when, in isolated instances, the term "air release agent" might be somewhat more appropriate.

Guide to Further Reading

Hess, M. *Anstrichmangel und Anstrichschaden, Ursachen und Verhütung*, 2. Aufl. (1954) Berliner Union, Stuttgart
Montle, J., Markowski, H., Lodewyck, P., Schneider, D. U.S. Patent 4323690 (1982)
Plueddemann, E., *Silane Coupling Agents* (1982) Plenum Press, New York
Plueddemann, E., Clark, H., U.S. Patent 3258477 (1966)
Plueddemann, E., *Polyelectrolytes and their Applications* (1975) D. Reidel Publishing, New York.
Stoffer, J., Montle, J., Somasiri, N., 4778 910 (1988) U.S. Patent
Scheiber, J. *Die Polymerisatharze* (Chemie und Technologie der Künstlichen Harze, Bd. I) 2. Aufl. (1961) Wissenschaftl. Verlag Stuttgart
Stoffer, J., Oostendorp, D., Proceedings of the Fifteenth Water-Borne and Higher-Solids Coatings Symposium, February (1988) p. 513
Vold, M.J., Vold, R.D., *Colloid Chemistry* (1964) Reinhold Publishing Co., New York
Wright, W.D., *The Measurement of Appearance* (1975) Van Nostrand Reinhold Co., New York
Zorll, U., Untersuchung zur Glanzmessung an Anstrichschichten mit dem Goniophotometer, 1/1963, Deutsche Farbenzeitschrift

17 The Performance Mechanisms of Defoamers and Air Release Agents

17.1 Determining the Rate of Bubble Rise

At the beginning of any comprehensive study of "defoaming" and "air release" phenomena, one should first closely examine *the correlation of bubble radius and system viscosity to "velocity of bubble rise"*; Figure 17.1 describes the interlocking relationships involved.

According to the above equation, *high viscosity systems* will reduce the rate of bubble rise; therefore, "air release" is the term which more appropriately describes the phenomenon at hand. As expected, *low viscosity systems* will allow accelerated bubble release rates—so the more germane descriptive term in the latter case is "defoaming". This means that "bubble removal" in low viscosity systems nearly always occurs *at or above* the coating surface (rather than beneath the surface as in high viscosity scenarios); furthermore, such removal is generally comparatively easy. In other words—the deeper that foam bubbles are trapped within the coating, the more difficult the removal process. As an additional corollary, smaller bubbles may cause more problems than larger ones.

Both defoamers and air release agents can be considered as *interfacially active substances which are designed to be selectively incompatible with the foam lamella.* In terms of distinguishing characteristics, defoamers are often siloxanes or related derivatives, whereas air release agents are lamella-disrupting polymers which may contain neither siloxanes nor related moieties. Siloxane products (with their tailored surface tension properties) are designed to function specifically at the coating surface and/or at all liquid/air interfaces; therefore, foam bubbles which have already risen through the bulk of the film volume can often be easily removed. Air release agents, in contrast to defoamers, generally display enhanced lamella-disrupting influences *within*

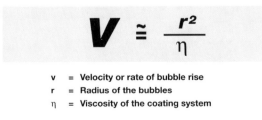

$$V \cong \frac{r^2}{\eta}$$

v = Velocity or rate of bubble rise
r = Radius of the bubbles
η = Viscosity of the coating system

Figure 17.1 Velocity of bubble rise

the coating film itself. Nevertheless, the mere disruption of foam bubbles is not always enough to release entrapped air. The missing link which can dramatically improve performance is often the inducement of increased bubble radius.

How can one increase bubble radius? The special polymers contained in air release agents are designed to *first* simultaneously contact multiple "average-radius" bubbles in localized areas beneath the coating surface; *next*, the very same polymers rapidly induce lamella/lamella contact between adjacent bubbles. *Finally*, the resultant interbubble contact points then rupture in a very special fashion that allows the gas from many small bubbles to flow into a "pooled area". The end result is quite simple—a series of relatively large-radius foam bubbles is formed. Often, radius length may increase by a factor of 15 to 20; on a practical basis, air release efficiency and effectiveness can increase very dramatically—by a factor as great as 15^2 to 20^2 (225 to 400)!

17.2 The Three Prerequisites of Performance Enhancement

Defoamers are low surface tension liquids which, in general, must demonstrate the following three properties:

A. Controlled insolubility and/or incompatibility in the medium to be defoamed
B. Positive entering coefficient
C. Positive spreading coefficient

When the entering coefficient is positive, the defoamer can, of course, enter the foam lamella. If, in addition, the spreading coefficient is positive, then the active ingredients of the defoamer can now actually spread on the interface (Fig. 17.2).

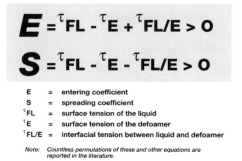

$$E = {}^\tau FL - {}^\tau E + {}^\tau FL/E > 0$$

$$S = {}^\tau FL - {}^\tau E - {}^\tau FL/E > 0$$

E	=	entering coefficient
S	=	spreading coefficient
${}^\tau FL$	=	surface tension of the liquid
${}^\tau E$	=	surface tension of the defoamer
${}^\tau FL/E$	=	interfacial tension between liquid and defoamer

Note: *Countless permutations of these and other equations are reported in the literature.*

Figure 17.2 Special coefficients

Defoaming Action

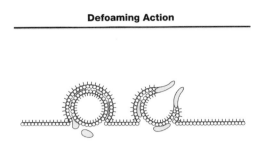

Figure 17.3 Mechanism

Because of this spreading effect, the foam stabilizing surfactants are pushed away, and the previously elastic, disturbance-resisting lamella is replaced by a lamella film which demonstrates both lower surface tension and reduced cohesive forces. This dynamic defoaming mechanism can be further accentuated (especially in aqueous systems) by the addition of finely dispersed hydrophobic particles. To a large extent, the defoamer liquid serves as a carrier medium which transports the particles into the foam lamella. On one hand, the hydrophobic particles themselves function in the hydrophilic liquid lamella as "foreign" particles, and therefore contribute to foam destabilization through the reduction of cohesive forces. On the other hand, such particles adsorb or "capture" surfactant molecules on their surfaces, thereby forcing the foam lamella to break.

With positive coefficients, the defoamer can enter the foam lamella and puncture the foam bubbles as shown in Fig. 17.3.

17.3 Defoamers for Environmentally Friendly Aqueous Systems

Many defoamers for aqueous systems can be classified into two main groups: *mineral oil defoamers* and *silicone defoamers*. Broadly speaking, mineral oil defoamers are intended primarily for usage in flat and semigloss emulsions and plasters; silicone defoamers are primarily for premium water-based industrial coatings.

17.3.1 Mineral Oil Products

A mineral oil defoamer is generally composed of approximately 80% carrier oil (often highly modified) and 15% hydrophobic particles. The remaining 5% are emulsifiers, biocides, and other enhancing ingredients.

As the carrier oil, aromatic or *aliphatic* mineral oils can be considered. *Aromatic* products, though, are seldom employed since they may cause premature yellowing of the paint film, and since they may present physiological handling risks due to their high levels of polycyclic aromatic hydrocarbons. Hydrophobic particles, as well as carrier oils, can exhibit a decisive influence upon defoamer behavior. Special fumed silicas (in addition to metal stearates and fatty acid derivatives) are generally employed. Newly patented defoamer structures greatly differentiate themselves from those of traditional defoamers, especially since the more modern products are often based upon technology which involves the utilization of specially designed polyurea compounds as the hydrophobic particles. In addition to demonstrating increased defoamer action, these patented compounds exhibit two additional advantages:

1. The polyurea is integrated in situ into the carrier oil, thus leading to a much finer distribution of the particles; this dramatically reduces separation tendency and also provides increased storage stability of the defoamer. (Please note that often carrier "oils" are so highly modified that one could quite easily consider the resultant moieties as rather "non-oil-like" in nature.)
2. Because of the larger specific interface available, the adsorption capacity in regard to surfactants is greater. This, in turn, assures optimal defoamer activity even after long-term storage of the finished coating.

In the presence of a positive spreading coefficient, the defoamer will then actively spread at the interface, thereby "mopping up" the surfactant molecules. This reduces the foam stabilizing effects of both electrostatic repulsion and Gibbs elasticity, thus leaving a much less stable lamella with lower surface tension and smaller cohesive forces. The effectivity of defoamers can be increased by incorporating finely dispersed hydrophobic particles which, when carried into the foam lamella, act as foreign bodies in the hydrophilic liquid, adsorb the surfactant molecules and thus destabilize the foam—thereby destroying the structural integrity of the foam bubbles. A recent development on this technology front is the use of aliphatic polyurea hydrophobes, and derivatives thereof, as shown in Figure 17.4.

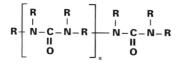

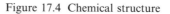
Figure 17.4 Chemical structure

The emulsifiers utilized in defoamers are necessary to disperse the particles within the carrier oil. In addition, they exert a positive influence upon defoamer incorporation ease. Where premium defoamers are required for high gloss emulsions, it is often desirable to augment performance by adding a small amount of silicone to increase the spontaneity of defoaming action.

17.3.2 Silicone Products

The second defoamer category for aqueous systems is silicones. Because of their special functionalities, silicone products are more expensive than mineral oil counterparts, and are therefore intended for usage in premium coatings formulations. For the most part, emulsions of strongly hydrophobic silicones (dimethylpolysiloxanes and/or polyether-modified dimethylpolysiloxanes derived from the base materials in the 5000 to 50000 mPas range displayed in Figure 17.5) are employed. (Please note that, in various circles of the industry, several widely accepted—and sometimes possibly quite different—"abbreviation shortcuts and variants" are employed. As mentioned in other chapters, the author has observed the occurrence of an absolute plethora of different spellings, punctuations, and unit-to-unit conversion rules for certain technical terms and chemical nomenclature expressions in the industry. As paradoxical as it may seem, often even the accepted industry reference books and dictionaries differ greatly on certain conversion rules, spelling issues, "pluraliza-tion conventions", hyphenation/punctuation rules and word-division protocols. Accordingly, several alternate conversion rules and/or grammatical usage prac-tices—all equally correct—may indeed be employed by some readers. This commentary applies, of course, not only for this chapter—but also for all other chapters in this textbook.)

Silicone defoamers can also be combined with hydrophobic particles (polyureas) to enhance silicone oil dispersibility and to improve defoaming performance. The primary advantage of silicone defoamers (as compared to mineral oil products) is that they neither reduce gloss in high gloss systems, nor do they alter color acceptance in pigment paste systems. The individual products vary according to not only the particular hydrophobic silicone oil employed, but also according to emulsifier type. Dependent upon which product is chosen, differences in crater susceptibility and in storage stability can be noted. However, in many instances, optimal incorporation of the defoamer—via higher shear forces—can help avoid deleterious side-effects, thereby accentuating performance enough to achieve completely crater-free coatings.

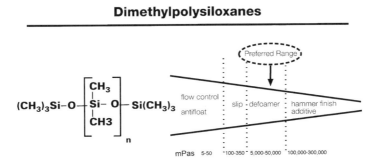

Figure 17.5 5000 to 50,000 mPas range

17.4 Defoamers for Environmentally Friendly Solvent-Based Systems

17.4.1 Silicone Products

For coatings systems which contain organic solvents, the previously described mineral oil defoamers are often not appropriate, especially since the spreading capacity of the mineral oil products may not be sufficient. Defoamer compounds with lower surface tensions are required, and polysiloxanes therefore play the dominant role as defoamer base substances. Because of their exceedingly positive spreading coefficients, polysiloxanes are the most dominant group of defoamers for solvent-based coatings.

The optimal defoamer choice involves choosing a product that can achieve enough balanced incompatibility to destroy foam, but not enough to cause craters. The compatibility of polysiloxanes, as mentioned in Part III, can be controlled and modified by the proper synthetic manipulation of organic side-chains. Polyether side-chains provide increased compatibility in polar systems. On the other hand, the use of a methylalkyl modification will increase surface tension and thereby reduce the foam stabilizing effect. One of the latest developments is fluorine-modified silicones. These "fluoro-silicone defoamers" demonstrate particularly low surface tensions and strong defoaming effects. Of course, defoamer moieties are not restricted to the aforementioned categories. For instance, silicone-free defoamers for solvent-based coatings are also available; they are normally polymeric in nature and defoam in direct relation to their controlled incompatibility.

Whenever one chooses a polysiloxane, chemical structure is indeed the decisive point. For example, a few of the relatively short-chain polysiloxanes may demonstrate foam-stabilizing, rather than foam-destabilizing (defoaming), behavior. Whether or not a particular polysiloxane functions as a foam stabilizer or as a defoamer depends upon both the polysiloxane's compatibility and its solubility in the liquid medium at hand; only selectively incompatible and insoluble polysiloxanes can successfully operate as defoamers. The controlling factor here is, in part, the molecular weight or chain length of the silicone. Quite unsurprisingly, the selection of the proper defoamer can be characterized, in practical terms, as a true "balance act" between compatibility and incompatibility. As illustrated in the figurative display of a "see-saw" in Figure 17.6, too much incompatibility leads to coating defects; however, too much compatibility leads to foam stabilization rather than to foam removal. The required balance and the resultant "selective incompatibility" can be achieved through a variety of the aforementioned silicone chemistry techniques.

17.4.2 Silicone-Free Polymeric Products

In regard to defoaming behavior in solvent-based systems, special polymeric products (operating through selective incompatibility) are often worthy of consideration. To

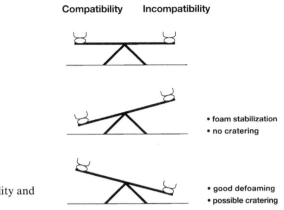

Compatibility   Incompatibility

• foam stabilization
• no cratering

Figure 17.6 Critical balance of compatibility and
incompatibility

• good defoaming
• possible cratering

achieve the proper balance between "compatibility" and "incompatibility", one can intentionally modify polymeric polarity, along with molecular weight and/or statistical aspects of molecular-weight distribution. As opposed to silicone defoamers, polymeric defoamers which are too compatible do not lead to foam stabilization. (Defoamer behavior is simply too weak or perhaps even non-existent.)

17.5 Selection Criteria, Test Methods, and Performance Measurement

Since defoaming action itself is the most important selection criterion, many practical test methods are employed. The fastest and easiest method is the evaluation of the defoamer in pure resin. In particular, air is incorporated into the resin as one observes how rapidly bubbles break (or more precisely, how quickly the foam volume is reduced). Nevertheless, an evaluation such as the above must be characterized as only a "pre-test" since the final coating formulation contains numerous additional components which can also influence foam behavior. It is therefore mandatory that one perform defoamer evaluations in the final coating system itself.

In order to gauge performance among, for instance, the control sample and several comparative samples (in either a "ladder-study" or in a defoamer test series), a reproducible method of incorporating air, or producing foam, is recommended. Because the objective is to achieve comparable results regardless of the methodology employed, then "absolute" methods are not of paramount importance. The main point is merely the production of as much or as little foam as necessary to differentiate the test results; the following methods can be employed:

- Shaking in a paint or laboratory shaker
- Stirring with a dissolver
- Incorporation of air via the back and forth movements of a perforated metal plate
- Introduction of air via pump
- Air incorporation with a brush
- Rolling with an open-pore roller

After air has been incorporated, then one can observe (1) the foam reduction over time, and/or (2) the visually perceptible defects in the final applied paint film. Evaluation techniques include, for example, microscope examinations or weathering tests. (In addition, there are dozens of ASTM/DIN/ISO and related techniques.) It is often helpful to apply the foamed paint to a glass plate or to a clear or translucent plastic sheet, thereby visually observing (with light passing through the substrate) the film. In highly filled systems (such as plasters), density evaluation may also be quite useful; the most dense sample naturally exhibits the best foam removal.

Since great variations exist in regard to foam behavior in different coatings systems, one cannot logically recommend any one standard method. The coating should be tested approximately 24h after defoamer incorporation. In addition, comprehensive evaluations after storage (since certain defoamers may lose effectiveness over time) should be completed. Significant performance differences can often be noted after only four weeks storage at 50°C. Because many defoamers contain hydrophobic particles, it is recommended that one also monitor the stability (separation tendency) of the defoamers themselves.

Of course, convenient and easy defoamer incorporation is an important decision criterion which should especially be considered whenever one must avoid variation in defoamer behavior during the production of different paint batches. Defoamers are optimally effective when they are insoluble (to a controlled extent) in the medium to be defoamed, and when they display a certain degree of selective incompatibility. The importance of the aforementioned balance must always be kept in mind; as a result, the monitoring of potential side-effects such as the following should be an integral element of one's decision-analysis program.

- Gloss decrease
- Haziness in clear systems
- Crater formation tendency
- Possible influence on intercoat adhesion

Because individual defoamer products may behave differently in regard to potential side-effects (and also in regard to the idiosyncrasies of the particular coating system at hand), one must vigilantly consider the above variables when making the defoamer choice. Whether or not a particular coating system exhibits problems is dependent not only upon the formulation itself, but also upon substrate conditions and other factors (such as exactly how and under what parameters defoamer incorporation occurs). One should always attempt to include the most important application parameters in the test program. Such inclusion provides the only guarantee of obtaining both definitive and practical results.

Guide to Further Reading

Alexander, A.E., Johnson, P., *Colloid Science* (1949) Oxford University Press, New York
Bikerman, J.J., *Foams* (1973) Springer Verlag, New York
Boys, C. V., *Soap Bubbles* (1959) Dover Publications, Inc., New York
Federation of Societies for Coatings Technology, (1978) *Pictorial Standards of Coatings Defects* Philadelphia, PA
Fischer, E.K., *Colloidal Dispersions* (1950) John Wiley and Sons, Inc., New York
Hess, M., *Paint Film Defects* (1965) John Wiley and Sons, Inc., New York
Kruyt, H. R., *Colloid Science* (1952) Oxford University Press, New York
Sheludko, A., *Colloid Chemistry* (1966) Elsevier Publishing Co., New York

Part V

Integrative Discussions of Performance and Economics

18   Newly Patented Developments in Complex and/or Multi-Layer Coatings

Given the challenging demands of certain complex and/or multi-layer systems, the modern formulator's objective is *the provision of a proper balance of features among primer, basecoat, and topcoat systems*. Of course, this means that not only *intra*coating features (such as optimized pigment particle spacing), but also *inter*coating features (such as substrate wetting and intercoat adhesion) must be managed in a proper fashion. Accordingly, new chemistries have been developed*—with newly patented advances occurring to simplify the task of formulating multi-layer systems. The aforementioned developments, as shown below, are discussed within an integrated practical and theoretical framework:

- Introduction: Pigments and force fields
- Pigment mapping concept
  - Enhancing performance in primer surfacers: Two new synergies
    - Acrylates and silicones
    - Wetting agents and pigments
- Advanced silicone surfactants for basecoats
- New techniques for controlling, measuring, and enhancing topcoat performance
- Conclusion: Optimized intra- and inter-coating features

18.1   Introduction: Pigments and Force Fields

Regardless of whether one is formulating primers, basecoats, or topcoats—energy is added to the system during the dispersing process. The intermediate result is an "energy-rich" system that may eventually attempt to transform itself into a less energy-rich system. Both *intra*coating features and *inter*coating features can be integrally affected. For instance, if pigment stabilization has not occurred, then the finely dispersed pigments may build uncontrolled flocculates, with the end result possibly being sub-optimal corrosion protection in primers; deleterious flooding and floating in basecoats; lowered intercoat adhesion between primers and basecoats; and reduced performance in topcoats/clearcoats.

---

*Some of the more specific advances, particularly in the UV/EB coatings arena, are covered in Appendix IV.

Dozens of additional performance features are affected by the presence of such flocculates. For instance—*light scattering and particle size distribution* can be altered in such a way that lack of reproducibility in regard to color strength, gloss, opacity (hiding power), and other macroscopic appearance parameters (including synergistic flow and levelling effects) may occur.

Flocculate size distribution variables are, in turn, integrally related to both electromagnetic (attractive) and electrostatic (usually repulsive) forces—*interpigment forces* which must be controlled as crucial elements of the "first line of flocculation defense." Additional defense elements include steric forces and special combination effects. Because of the often pairwise occurrence of electromagnetic and electrostatic forces, one cannot adequately discuss one force type without at least introducing the other; however, the prime point of importance to the coatings formulator is generally the unequivocal *dominance* of the repulsive (dispersion-stabilizing) forces.

Although a comprehensive array (several dozen, to be exact) of mathematical formulas has been developed to describe nearly all formulator-influenced factors, the objective of this chapter is exclusively the presentation of practice-oriented methodologies to aid in the actual day-to-day formulation of coatings. Consequently, no attempt whatsoever at a comprehensive review of the previously discussed (please see, for instance, Chapter 3) molecular attraction and repulsion potentials will be made; quite the contrary, *only the practical end result will be presented in a synopsis diagram.* In terms of optimal formulator effectiveness, one merely has to shift the balance of attractive forces (V_A) and repulsive forces (V_R) so that an overall preponderance of repulsive forces occurs, as demonstrated in the shaded portion of Figure 18.1. What does this mean from a practical perspective though? How can one use information about force fields to improve performance? The practical answers to these questions serve as the focal point of this publication. For instance, from the chemist's standpoint, the aforementioned diagram can be employed not only to gauge overall coating performance, but most importantly, to design enhanced additives

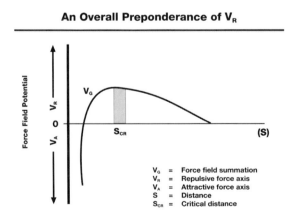

Figure 18.1 The importance of V_R

which can exhibit new synergistic effects in multi-layer coating systems. An overview of the end result—day-to-day performance improvement in flow, levelling, and pigment wetting—is provided in Sections 18.2 to 18.4.

18.2 The Pigment Mapping Concept

Background: One of the most powerful performance improvement concepts involves "pigment mapping", a procedure which allows complex formulations to be adjusted and fine-tuned within the framework of a multi-dimensional grid. A sampling of three mapping variables, expressed as either "grid axes" or specially demarcated "grid zones", are shown below:

- Pigment polarity ("X" axis)
- Resin polarity ("Y" axis)
- Zones of "additive compatibility" (Such zones are often denoted by superimposed rectangles and/or other two-dimensional circles, polygons, etc. Of course, more complex scenarios may involve special zones with three-plus dimensions.)

Even though selected aspects of the "pigment mapping concept" were briefly touched upon in Chapter 1, only now—near the conclusion of this text—can the true utility and potential of concepts such as pigment mapping come to the forefront. For instance, utilizing the "Surface Polarity of Pigments" diagram (Fig. 18.2)—one can, first of all, "map" various pigments; next, one can successively incorporate

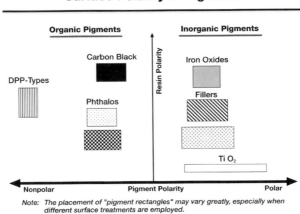

Figure 18.2 Surface polarity of pigments

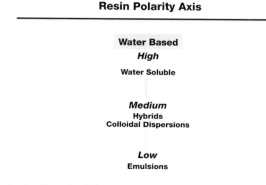

Figure 18.3 Resin polarity (Part 1 of 2)

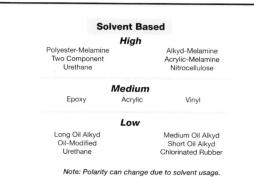

Figure 18.4 Resin polarity (Part 2 of 2)

information about resin polarity (please see Figures 18.3 and 18.4) into the overall grid concept. The next (and final) step is the superimposing of various "Additive Compatibility Zones" onto the grid. *What is the utility of this? What exactly can be gained from superimposing "Additive Zones" onto a grid already demarcated by zones of defined pigment and resin features? In a nutshell, one can easily determine exactly which additives match any particular combination of pigment and resin!*

Obviously, this particular concept can be further extended to include additional "intersecting planes" which represent, for instance, solvent features, substrate parameters, etc. In fact, one could ultimately construct a densely packed multi-dimensional grid, composed of hundreds of intersecting sub-components, to mimic nearly all of the variables and constraints encountered. Thus far, grids composed of several dozen planes and axes have already been constructed, but the current limitations of computer modeling have not stretched far enough to accommodate grids containing several hundred axes. Subsequent volumes of this textbook series will cover broader aspects of this most exciting endeavor to map and model (in an analogous fashion to the "human genome project") all germane features of coating performance.

The end result is, at least in well-developed studies, literally a series of several thousand grids, of which only six randomly chosen examples are shown (Figs. 18.5 to 18.10). Please note that the necessity for such a large number of grids is based on a variety of factors, one of which being the myriad influences of pigment treatments on the exact placement of "pigment blocks" on the grid surface.

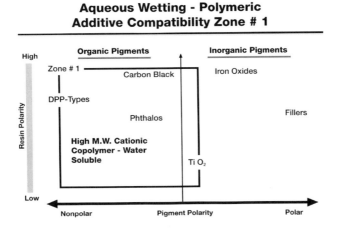

Figure 18.5 Zone 1

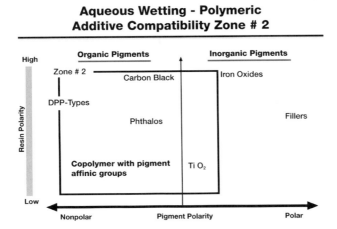

Figure 18.6 Zone 2

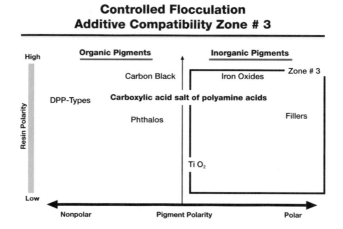

Figure 18.7 Zone 3

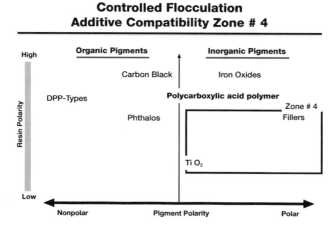

Figure 18.8 Zone 4

18.3 Enhancing Performance in Primer Surfacers: Two New Synergies

Primer surfacers for multi-layer OEM (original equipment manufacturer) and related applications are formulated to strict performance requirements in regard to gloss, levelling, circulation pipe stability (no settling or color drift), and the prevention of

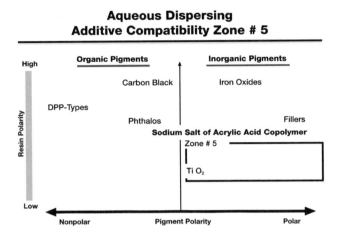

Figure 18.9 Zone 5

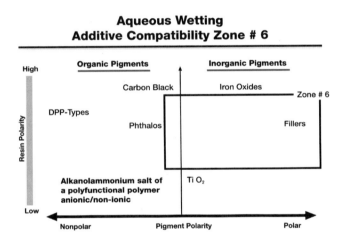

Figure 18.10 Zone 6

"ghosting" in the topcoat. Certain performance parameters, such as pigment behavior, can be properly controlled by wetting and dispersing additives. On the other hand, parameters such as surface structure and ghosting in the topcoat require flow and levelling additives. Within the context of improving performance, two new synergies, as shown, are overviewed:

SYNERGY #1: Wetting Agents and Pigments
SYNERGY #2: Acrylates and Silicones

18.3.1 Wetting Agents and Pigments

18.3.1.1 Deflocculation Versus Controlled Flocculation

Optimal gloss, Newtonian rheology, and resistance to color drift are achieved when all pigments and extenders are completely *deflocculated* by wetting and dispersing additives. In contrast, circulation pipe stability is best achieved by the *controlled flocculation* of pigments and extenders, or by the synergistic usage of rheology modifiers.

18.3.1.2 Corrosion Resistance: A New Role for Wetting and Dispersing Additives

To reconcile the opposing requirements of deflocculation and controlled flocculation, it is often essential to select a wetting and dispersing additive which provides a synergistic compromise. Selection depends mainly on the type of pigment used. For primer surfacers containing only inorganic pigments, controlled flocculation additives can be employed. On the other hand, for systems involving inorganic/organic combinations (including carbon black), deflocculating polymeric products have often proven most successful.

   Interestingly enough, one of the rather paradoxical advantages of certain controlled flocculation additives is not only proper pigment particle spacing, but also improved corrosion resistance—all achieved by special wetting and dispersing agent synergies. But can mere wetting and dispersing additives really provide *significant* corrosion resistance improvement? Yes, they can, sometimes up to seven points improvement on a ten-point scale (ASTM D-1654-79).

   How is this possible, and what exactly is the performance mechanism involved? From an overview perspective, the mechanism responsible for such dramatic results is twofold in nature—first of all, novel additives composed of enhanced polycarboxylic acid structures produce beneficial states of controlled flocculation. Secondly, these states of controlled flocculation are accompanied by the formation of uniquely structured pigment/resin/additive complexes (Fig. 18.11).

18.3.1.3 "Micro-Zones" and Physicochemical Changes

From the formulator's perspective, wetting and dispersing additives may seem out of place when cast in the role of corrosion-resistance enhancers, so what exactly lies at the heart of the aforementioned bifunctional performance mechanism? In other words, what is so unique about controlled flocculation and the so-called "pigment/resin/additive complexes"? These questions can be addressed only by overviewing the physicochemical changes (as shown below) which occur in three distinct "micro-zones" of properly enhanced coating films:

Incorporation of W & D Additive into Primers

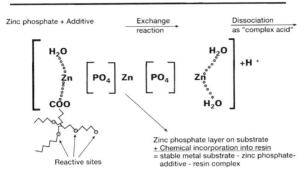

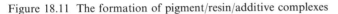

Figure 18.11 The formation of pigment/resin/additive complexes

Physicochemical Changes
Micro-Zone #1. Improved film surface formation: fewer crevices for corrosion and/or environmental contaminants (H$_2$O, salt, etc.) to enter
Micro-Zone #2. Improved pigment packing arrangements
Micro-Zone #3. Increased substrate adhesion because of the metal-affinity of the resultant pigment/resin/additive complexes
Definition of the three aforementioned "Micro-Zones"
#1. Air/Coating Interface
#2. Intracoating Interface (includes special "packing geometries")
#3. Coating/Substrate Interface

Of course, an exhaustive enumeration of the many factors contributing to micro-zonal behavior and corrosion resistance improvement is well beyond the scope of this synopsis—but the utility of novel polycarboxylic structures remains unmistakably clear: The results of repeated studies (utilizing over 19 additives in a battery of ASTM-based undercut corrosion and surface blistering evaluations: performance enhancement from 2 to 7 points on D-1654-79 and other tests) have demonstrated that—contrary to popular belief—corrosion resistance improvement is no longer the exclusive domain of specially designed inhibitors and anti-corrosion pigments.

Structurally speaking, the additives that can improve the corrosion resistance of primers and/or topcoats often contain specially modified polycarboxylic acids with multifunctional sub-components (e.g., carbon–carbon double bonds, carboxylic acid groups, amide structures, etc.). Because of these structures, the additives are easily incorporated into the resin system (as shown in the aforementioned diagram); accordingly, a very stable metal substrate/pigment/additive/resin complex is formed. Even in situations where a metal substrate is not employed—or even where corrosion resistance is only of marginal importance—one can still enhance performance in two out of three micro-zones.

Obviously, this does not imply that one should expect wetting and dispersing additives to replace anti-corrosion ingredients; nor does this imply that panaceas exist, or that all wetting and dispersing additives are capable of improving corrosion resistance. Statistically speaking, only 21% of the additives tested were successful in the corrosion enhancement arena; nevertheless, what this means to the formulator is that the judicious choice of wetting and dispersing additive is a vitally important tool in combating corrosion—a tool that is often underestimated and overlooked.

18.3.2 Acrylates and Silicones

To achieve proper levelling and to avoid Bénard cell formation during the drying process, additives can be used to calm the surface and to control surface flow. Such additives should not be susceptible to "ghosting" through the topcoat; as a result, acrylic levelling additives are excellent candidate products for this application. *From the formulator's perspective, the most optimal acceptance of the topcoat can often be achieved through the utilization of special acrylates and/or silicones which are designed to exhibit analogous backbone structure modifications.* (Example chemical structures are reviewed in Fig. 18.12.)

The multiple performance improvements which accrue from the usage of the aforementioned silicone/acrylate synergies include *(1) the ability to reduce surface tension to the 22 to 25 mN/m level (thereby matching the performance of certain fluoro-based products—without the high cost and foam stabilization accompanying the usage of some products in the latter category), while simultaneously providing perfect (2) flow, (3) levelling, and (4) intercoat properties.*

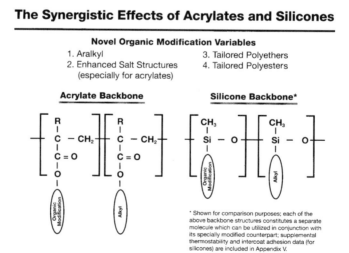

Figure 18.12 Synergy

18.4 Advanced Silicone Surfactants for Substrate Wetting in Environmentally Friendly Basecoats

Certain conventional organic surfactants (either ionic or non-ionic) do not always function optimally in environmentally friendly formulations as substrate wetters—primarily because of the relatively high usage levels required, and also because of possible negative influences on humidity resistance (obviously a side-effect of usage at elevated levels). Some surfactants, based partially on hydrophilic and hydrophobic–oleophilic structural units, orient to nearly all paint interfaces, not only to the air and substrate interfaces, but also to the pigment/resin interfaces. Advanced surfactants (where a hydrophilic block of tailored size and orientation is attached to a specially modified hydrophobic–oleophilic unit) tend to preferentially migrate to the air and/or substrate interfaces (variable migration according to structure), not to the pigment/resin interfaces. As a result, these additives are optimally effective in the control of surface tension and substrate wetting. The hydrophobic–oleophilic structural unit is usually either a dimethyl polysiloxane moiety or a fluorinated (and specially modified) aliphatic block.

In terms of a comparative overview of substrate wetters, the evolutionary advance of three different structural moieties (with special emphasis on products designed for environmentally friendly basecoats) is shown in Figure 18.13.

The most evolved structure, that of certain advanced silicone surfactants, provides not only *"wave-free" (see the case study in Figure 18.14) surface tension reduction down to the 22 to 25 mN/m range*, but also provides—especially in comparison to certain fluoro-based counterparts—*the added advantages of significantly lower foam (up to*

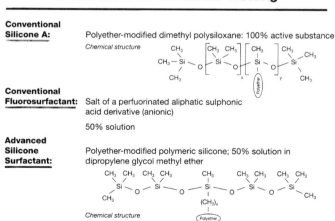

Figure 18.13 Evolutionary advance

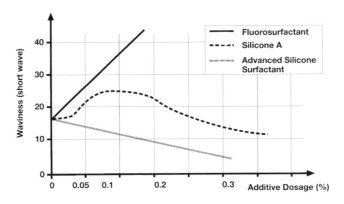

Figure 18.14 Case study: Comparative study of flow

90% less) and lower cost (in multi-layer systems, up to 70–95% less cost than with comparable fluoro-based products).

In illustration of the "wave-free", improved levelling properties that result from the usage of advanced silicone surfactants, the following abbreviated case study is offered: When utilizing various dosages of three representative additives in a black waterborne polyurethane-melamine baking automotive basecoat, an interesting surface flow development was observed. *Only the usage of an advanced silicone surfactant resulted in optimized surface flow.* In contrast, conventional silicones and fluorosurfactants were incapable of eliminating certain imperfections. For instance, at low usage levels, one conventional silicone actually increased waviness and decreased surface flow; however, at high levels, the waviness eventually displayed improvement. In contrast, the fluorosurfactant caused very strong orange peel at both normal and high dosages. For these investigations, 45 micron films were applied on glass panels with an automatic spray gun.

18.5 New Techniques for Controlling, Measuring, and Enhancing Topcoat Performance

Gloss, flow, surface slip, and hue stability are all important properties in topcoat systems, especially when challenging substrates and/or primer systems are employed. Such properties are integrally related to the optical characteristics and particle size parameters of the pigments. In OEM and other high performance topcoats, pigments are most economically used when optical performance is properly developed (this is

**The Physics of Performance Improvement
Through Particle Size Manipulation**

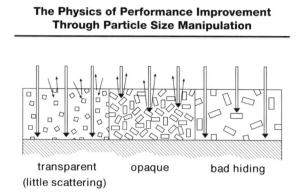

transpareut
(little scattering)   opaque   bad hiding

Figure 18.15 Particle size manipulation

generally the case when the pigments are deflocculated). As mentioned previously, hiding power and/or transparency can be tailored (in part) to optimize performance requirements through the proper manipulation of particle size parameters; a demonstration of the physics of such performance enhancement is shown in Figure 18.15.

Of course, the initial choice of pigment itself is quite important; nevertheless, as described previously, the selection of a tailored polymeric wetting and dispersing additive is sometimes even more critical. Ideally, one should combine the physical advantages accrued by particle size manipulation (as shown in Figure 18.15) with the chemical advantages offered by the advanced chemistry shown in Figure 18.16.

**Advanced Chemical Structure of a Topcoat
Additive Designed to Simultaneously Deflocculate
Organic and *Inorganic* Pigments**

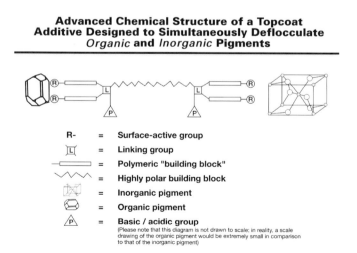

R-	=	Surface-active group
⊥	=	Linking group
▭	=	Polymeric "building block"
⟋⟍	=	Highly polar building block
▨	=	Inorganic pigment
⬡	=	Organic pigment
△P	=	Basic / acidic group

(Please note that this diagram is not drawn to scale; in reality, a scale drawing of the organic pigment would be extremely small in comparison to that of the inorganic pigment)

Figure 18.16 Advanced structure

One important advantage offered by topcoat additives designed to simultaneously deflocculate organic and inorganic pigments is the *absolute and total control of all pigment particle parameters*—via two mechanisms:

1. Steric hindrance
2. Electrostatic charge repulsion

Multi-layer systems which entail the simultaneous usage of primer and topcoat systems are often especially susceptible to problematic pigment states. For example, if *traditional* wetting and dispersing additives are utilized, then gloss-destroying flocculates may tend to form in certain topcoats because of the following two reasons: (1) improper surface parameters in the underlying primer, or (2) problematic interpigment spacing in the topcoat (provided that, of course, the topcoat is not a clearcoat). Please note that, semantically speaking, the term "topcoat" can obviously subsume all "clearcoats", but for the purposes of the isolated example above, only pigmented topcoats are considered.

In contrast, *advanced* wetting and dispersing additives which utilize the chemistry displayed in the previous diagram alleviate problems such as those described above. Although an enumeration of the full laboratory results of vast empirical studies is certainly beyond the scope of this overview publication, certain "practical end results" are definitely worth mentioning. For instance, as shown in the previous diagram, the special combination of electrostatic and steric hindrance effects imparted by the "R" groups (surface-active), the "P" groups (basic/acidic), and the "L" groups (linking) is more than capable of fulfilling the initial force field requirements that were set forth in Section 18.1. What does this mean, though, from a practical perspective? What benefits accrue to the formulator? *In a nutshell, virtually all deleterious forms of flocculation are prevented.*

18.6 Conclusion: Optimized Intra- and Inter-Coating Features

Primers, basecoats, and topcoats pose multiple challenges in regard to color, gloss, opacity, corrosion protection, flooding and floating, intercoat adhesion, and optical performance. The proper control of a variety of physicochemical variables can simplify the process of formulating defect-free coatings. In particular, the following principles must be observed:

1. Before *inter*coating problems can be properly controlled, *intra*coating phenomena must first be addressed.
2. Chemical synergies in the following areas can greatly enhance performance in primer-surfacers:

   - Acrylates and silicones
   - Wetting agents and pigments

Major improvements (in regard to flow, levelling, pigment control, and corrosion resistance) can occur.

3. Advanced silicone surfactants often provide superior performance when compared to either conventional silicones or to fluorosurfactants. In addition to achieving improved surface flow, one can lower cost and reduce foam—especially in comparison to formulations containing fluorosurfactant analogs.

4. New techniques for controlling and enhancing topcoat performance exist. The tailored utilization of the following chemical moieties results in complete control of interparticle parameters:

- "R" groups (surface-active)
- "P" groups (basic/acidic)
- "L" groups (linking)

The above moieties are integrated into a high molecular-weight polymeric backbone with tailored (often highly polar) building blocks. Furthermore, wetting and dispersing additives which utilize the above chemistry not only improve performance, but they also prove amenable to the enhanced methods of pigment fine-tuning and control described herein.

Guide to Further Reading

Degussa Corporate Communication, *Bewitterungs- und Lichtechtheits-Prüfungen an mit Ruß pigmentierten Grau- und Schwarzlacken*, Schriftenreihe Pigmente Nr. 22 (1968) Frankfurt/M
DIN 53902, Teil 1, 3 (1984) *Bestimmung des Schaumvermögens, Lochscheiben Schlagverfahren*; DIN 53902, Teil 2, 12 (1977) *Bestimmung des Schaumvermögens, modifiziertes Ross-Miles-Verfahren*
EPS 0115585; USP 4696761; USP 4314924
Ferch, H., Kittel, H., *Lehrbuch der Lacke und Beschichtungen* (1974) Verlag W.A. Colomb, Berlin-Oberschwandorf, Band 2
Ferch, H., Lambermont, M., Oelmüller, R., *Adhäsion* (1985) 7/8, S. 21–24
Gress, W., *Polymers Paint Colour* J. (1984) Nr. 4122, S. 452–466
Honak. E.R., Martinez, P., Farbe + Lack (1970) 76, p. 62
Koch, W., *Oberfläche—Surface 22* (1981) 107
Kockott, D., Fatipec-Kongreßbuch I (1984) 435
Kulkarni, R.D., Goddard, E.D., Kanner, B., *Ind. Eng. Chem. Fundamentals* (1977) 16(4), S. 472-474
Lacke + Farben AG and NOF Corporation, Patent # 0 775 175A, BASF, Storage-Stable Aqueous Compositions
Lenhard, W., Report of Speiss Hecker Meeting: Future Paint Technology (in German), *Farbenkreis* (1996) 11, 10–11
Noack, S., *Statistische Auswertung von Meß- und Versuchsdaten mit Taschenrechner und Tischcomputer*, (1980) De Gruyter, Berlin
Orr, E.W., *Corrosion-Improvement Factors for Direct-to-Metal and Primer Coatings*, 1995-1996 Lectures Available to FSCT Constituent Societies (special monograph series in support of FSCT lectures)
Orr, E.W., *Proceedings of the Association of Formulation Chemistry, Las Vegas*, September 2–4 (1997)
Papenroth, W., Koxholt, P., *Farbe + Lack* (1976) 82, p. 1011

Rechmann, H., Epple. R., Rosendahl. F., Schwindt, R., Vial, F., Fatipec-Kongreßbuch X (1970) 3

Ross, S., Nishioka, G., *J. Colloid Interface Sci.* (1978) 65 (2), S. 216–224

Sato, T.J., *Physical-Chemical Properties of Suspensions: Effect Of Titanium Dioxide Concentration On Physical Properties of Titanium Dioxide Suspensions*, Japan Soc. Col. Mat. (1996) 69(10), p. 699–707

19 The Integration of Technical and Economic Considerations

19.1 Introduction

As presented in Chapters 1 to 18 (and in the appendices), numerous case studies have empirically demonstrated the magnitude of performance improvement achieved by the enhancement of (1) wetting and dispersing, (2) interfacial tension, and (3) defoaming. For instance, new polymeric wetting and dispersing agents can improve not only color reproducibility, but they can also enhance pigment utilization in many systems. Instances involving up to 60–70% *less pigment utilization* (for equivalent color development) are not unusual. Concomitant economic advantages of similar magnitude can also be obtained with the interfacially active and/or defoaming products discussed.

The integration of technical and economic considerations, although of paramount importance, is often neglected in the hectic problem-solving rush of everyday life. But the race belongs to the winners—and to those who successfully intermesh technology and good business sense. Consistent with the dual theoretical and practical approaches of this publication, concrete measures for achieving "techno-financial" integration have been tabulated in a comprehensive series of easy-to-use charts, diagrams, and business/marketing forms. Several dozen formats*, all addressing different technical and economic considerations—ranging from PVC (pigment-volume concentration) to NPV (net present value of the performance improvement from a financial standpoint)—have been developed. Simultaneous display of all formats (there are over 80 in merely the first phase) is certainly beyond the scope of this particular publication; however, a sampling of the more basic "starter formats" is definitely worth perusal.

19.2 How to Measure the Economics of Coatings Reengineering

To appropriately accent the integrated nature of the above formats, a special exhibit entitled *"Reengineering Overview Chart"* has been prepared with the express purpose

---

*Semantically speaking, for the purposes of this particular publication, the term "format" (derived from the computer-generated nature of the forms) subsumes all the aforementioned charts, diagrams, and forms.

Table 19.1   Reengineering Overview Chart

- ◆ Financial aspects of reengineering (What effect will additive usage and/or reengineering changes have upon capital utilization, sales growth, sales price, asset utilization, stock price, and other factors?)

    - Profitability ratios
    - Growth ratios
    - Credit ratios
    - Stock ratios
    - Inventory and/or depreciation ratios

- ◆ Preliminary reengineering worksheet: Initial economic considerations from the salesperson's perspective

    - Who? What? When? Where? Why?
    - Required technical information

- ◆ Preliminary reengineering worksheet: Initial technical considerations

    - Formulation
    - Application and related details
    - Desired performance changes
    - Logistics of a potential project

of previewing the "starter formats". Accordingly, portions of the "Overview"—along with an array of germane coatings industry financial calculations—are displayed in Tables 19.1 to 19.4, whereas two additional formats, chosen at random, are displayed in Tables 19.5 to 19.6. (As always, please note that considerable variability—in regard to abbreviations, semantics, and nomenclature—exists throughout the industry. Numerous permutations of charts and equations—all equally correct—can be utilized.)

Absolutely no claim to "patented uniqueness" or comprehensiveness applies to the formats themselves; quite the contrary, the charts represent merely a concise, organized tabulation of key issues and questions with which management and/or technical personnel may possibly find themselves confronted every day. The utility of the formats* is simply that of synthesizing both economic and technical aspects into an integrated whole. Without doubt, the parameters and variables displayed are

---

*Obviously, the formats displayed are—in and of themselves—rather situation-specific. Each financial analyst, sales manager, salesperson, plant manager, formulator, and chemist will undoubtedly need to employ different "question-and-answer scenarios" to appropriately define the actual conditions which he himself routinely encounters. By no means whatsoever is it necessary for one to employ all 80 first-phase formats though. In practice. two or three charts (or even sub-charts) may prove sufficient for gauging the success of a typical reengineering endeavor.

Table 19.2   Financial Aspects of Reengineering: Part 1 of 3 (What effect will additive usage and/or reengineering changes have upon capital utilization, sales growth, sales price, asset utilization, stock price, and other factors?)

A. Profitability ratios*

1. Capital turnover	Total sales after reengineering
	Tangible assets—short-term accrued payables
2. Return on capital	Net income after reengineering + minority interest + tax adjusted interest
	Tangible assets—short-term accrued payables
3. Earnings margin	Net income after reengineering + minority interest + tax adjusted interest
	Sales
4. Return on capital before depreciation	Net income after reengineering + minority interest + tax adjusted interest + depreciation (a.d.1)
	Tangible assets—short term accrued payables

B. Growth Ratios

1. Growth of total return	Net amount earned for total capital in post-reengineering time period
	Net amount earned for total capital in base (or previous) time period
2. Sales growth	Sales in post-reengineering time period
	Sales in base (or previous) time period

*Of course, all variables and factors in these ratios should incorporate "post-reengineering" changes; this assures accurate measurement of enhanced profitability, growth, etc.

merely the tip of the iceberg; numerous augmenting factors (including perhaps the company-specific idiosyncrasies of production, finance, customer relations, transport, quality assurance, environmental regulations, etc.) may be required in many situations. Extensive tools in the project management arena—such as Gantt charts, PERT ("Program Evaluation and Review Technique") charts, and specialized Go/No-go templates form the crux of many formats, especially in the second and third phases of reengineering analysis. In addition, the triple scientific disciplines of "decision analysis", "problem analysis", and "statistical analysis" nearly always prove essential to proactively enhancing performance and/or to solving the problem at hand.

Table 19.3   Financial Aspects of Reengineering: Part 2 of 3 (What effect will additive usage and/or reengineering changes have upon capital utilization, sales growth, sales price, asset utilization, stock price, and other factors?)

C. Credit ratios

1. Post-reengineering "quick ratio"

$$\frac{\text{Current assets—inventories}}{\text{Current liabilities}}$$

2. Post-reengineering "current ratio"

$$\frac{\text{Current assets}}{\text{Current liabilities}}$$

3. Post-reengineering cash ratio

$$\frac{\text{Cash or cash-equivalent items}}{\text{Current liabilities}}$$

4. Post-reengineering cash flow to total capital invested

$$\frac{\text{Cash flow from post-tax operations + tax-adjusted interest from all sources}}{\text{Tangible assets—accrued payables}}$$

D. Stock ratios

1. Post-reengineering price/earnings ratio

$$\frac{\text{Share price}}{\text{Earnings per share}}$$

2. Post-reengineering sales per share

$$\frac{\text{Sales}}{\text{Weighted average of the outstanding common shares}}$$

3. Post-reengineering sales per dollar of common stock prorated at current market values

$$\frac{\text{Sales}}{\text{Weighted average of the outstanding shares X stock price}}$$

Table 19.4   Financial Aspects of Reengineering: Part 3 of 3 (What effect will additive usage and/or reengineering changes have upon capital utilization, sales growth, sales price, asset utilization, stock price, and other factors?)

E. Standard inventory and/or depreciation ratios

1. Post-reengineering inventory turnover

$$\frac{\text{Cost of goods sold}}{\text{Inventory (including possible LIFO reserves)}}$$

2. Post-reengineering depreciation to sales

$$\frac{\text{Depreciation expense}}{\text{Sales}}$$

3. Post-reengineering depreciation to gross plant

$$\frac{\text{Depreciation expense}}{\text{Gross plant}}$$

Table 19.5   Preliminary Reengineering Worksheet: Initial Economic Considerations from the Salesperson's Perspective

I. Who? What? When? Where? Why?

Salesperson, company, contact, phone	Customer need	Volume potential	Prob. of sale (%)	Timing

II. Required technical information
   (System? Performance? Technology?)

   A. Paint system (Check two)

[Solvent W–R Latex]	[Air dry Bake Other]

   B. Performance desired (brief description)

   C. End-use application (brief description)

19.3 Conclusions

The reengineering of coatings performance presents a series of challenges that can be mastered. Successful reengineering allows one to conquer the dichotomy between the technical and financial objectives of the coatings firm. Novel chemical structures can be utilized in conjunction with innovative formulating procedures to provide cost-effective, high performance solutions. In particular, specially designed additives can facilitate the achievement of the following three objectives:

- The reengineering of wetting and dispersing properties
- The enhancement of interfacial tension parameters
- The alleviation of foam

Within the context of achieving the above objectives, one automatically attains an economically efficient operating state in which both technical and economic factors assure the continued viability of the firm and its profit mission.

Table 19.6 Preliminary Reengineering Worksheet: Initial Technical Considerations

Formulation

1. Exactly what proportions of what must be added to what? (Attach product use sheet, if available)

   _____

2. How is the formulation prepared?

   ☐ Dissolver ☐ Ball mill ☐ Three-roll mill ☐ Sand mill ☐ Shaker ☐ Other _____

   Procedure _____

   _____

Application + related details

1. ☐ Spray ☐ Conventional ☐ Airless ☐ Brush ☐ Roller ☐ Pour ☐ Dip
   ☐ Other _____

2. Describe substrate or undercoat _____

3. Details: Temperature _____ Thickness _____ Pressure _____ Nozzle _____
   Other details: _____

4. Additives previously evaluated _____

Desired Performance Changes

☐ Flooding/floating	☐ Foam/blisters	☐ Viscosity stability	☐ Adhesion
☐ Sedimentation	☐ Levelling/flow	☐ Hiding power	☐ Salt
☐ Spray	☐ Wetting	☐ Sagging	☐ Weathering
☐ Gloss	☐ Conductivity	☐ Cratering	☐ Other

Logistics of a Potential Project

1. What materials would be required to formulate and test this formulation?

   a. Paint raw materials: _____

   b. Special substrates/test equipment, etc. _____

Guide to Further Reading

Barron, W.M., Dielmann, H. J., *Prozessführung in den USA* (1986), German American Chamber of Commerce, New York

Dowd, E.J., *Air Pollution Control Engineering and Cost Study of the Paint and Varnish Industry* (1974), Office of Air Quality Planning and Standards, Research Triangle Park, NC

Grafer, H., *Annual Report—Der US-amerikanische Jahresabschluß* (1992) Schäffer-Poeschel Verlag, Stuttgart

Kittel, H. *Farben-, Lack- und Kunststofflexikon, Nachschlagewerk über Begriffe, Verfahren, Rohstoffe, Pruefmethoden, Zwischenprodukte und Fertigungserzeugnisse* (1952) Wissenschaft/Verlagsgesellschaft, Stuttgart

Kittel, H., *Pigmente, Herstellung, Eigenschaften, Anwendungen* (1960) Stuttgart Wissenschaft/Verlagsgesellschaft, (related r. m., *Umweltschutztechnik—Eine Einführung*, 3. Auflage (1992) Springer-Verlag, Berlin; *Umwelt und Betrieb—Umweltrecht für die betriebliche Praxis* (1990) Erich Schmidt Verlag)

Olfert, Kfm. K. *Finanzierung*, Ludwigshafen (Rhein) (1992) Kiehl

Statsky, W.P., Hussey, B.L., Diamond, M.R., Nakamura, R.H., *West's Legal Desk Reference* (1991) West Publishing Company, St. Paul, MN

Stück, H.-H., *Grundwissen Steuern* (1993) Wilhelm Heyne Verlag GmbH, Munich

Appendices

Appendix I: 200 Starting Formulations

This appendix describes 200 formulations, divided into approximately 20 "formulation sets" (Tables I.1 to I.20). All information and data stated herein, although in no way guaranteed, are based upon tests and reports considered to be reliable and are believed to be accurate. No warranty, either express or implied, is made or intended. Use should be based upon one's own investigations and appraisals. Any recommendation should not be construed as an invitation to use a material in infringement of patents. Neither the author nor the publisher accepts responsibility for printing errors. In addition, neither the inclusion nor the exclusion of various products is indicative of endorsement or non-endorsement by the author or publisher. Consistent with standard reference book publishing conventions, any and all trade names are the respective properties of their owner(s); no entry in this reference work should be regarded as affecting the validity of any trademark or service mark; accordingly, trademark symbols are not used. Please note that certain information—in the body and/or appendices of this textbook—is based on data reported by outside parties (for instance, by chemical suppliers, distributors, etc.) In addition, the author has observed the occurrence of an absolute plethora of different spellings, punctuations, and unit-to-unit conversion rules for certain technical terms and chemical nomenclature expressions in the industry. As paradoxical as it may seem, often even the accepted industry reference books and dictionaries differ greatly on certain conversion rules, spelling issues, "pluralization conventions", abbreviation shortcuts, hyphenation/punctuation rules and word-division protocols. Accordingly, several alternate conversion rules and/or grammatical usage practices—all equally correct—may indeed be employed by some readers. The aforementioned statements apply not only to the appendices, but also to all other materials in this textbook.

Because of the compact nature of Appendices I to V, please note that an intentional departure from the "chapter outline system" is employed on all subsequent pages. Owing to space limitations, only a representative cross-section of chapter-specific additives is included in the appendices (Figs. I.1 to I.3; Table I.21 to I.23). Because of standard publishing industry conventions, some alphabetic and/or numeric codes were required to be used, in some cases, for multiple references. Accordingly, in situations where the reader may be interested in augmented information, please contact the publisher or author.

Table I.1 High Quality Industrial Coatings: Formulation No. 1. Starting Formulations for Pigment Concentrates with Acrylic Resin and Polymeric Additive A

	Titanium dioxide PW 6	Organic yellow PY 151	Iron oxide yellow PY 42	Chrome yellow PY 34	Organic red PR 170	Molybdate orange PR 104	Iron oxide red PR 101	Organic red–violet PV 23	Phthalo-cyanine green PG 36	Phthalo-cyanine blue PB 15	Carbon black PBk 7
Acrylic resin*	26.0	30.0	27.0	24.8	5.0	23.7	19.3	45.0	30.0	45.0	30.0
Methoxypropylacetate	5.4	11.7	4.8	0.0	8.3	0.0	0.0	21.7	24.0	21.7	14.0
Polymeric additive A** (30%)	3.3	23.3	9.7	10.7	26.7	10.8	15.2	13.3	21.0	13.3	28.0
Aerosil 200	0.3	0.0	0.5	0.5	0.0	0.5	0.5	0.0	0.0	0.0	0.0
Pigment	65.0	35.0	58.0	64.0	40.0	65.0	65.0	20.0	25.0	20.0	28.0
	100.0	**100.0**	**100.0**	**100.0**	**100.0**	**100.0**	**100.0**	**100.0**	**100.0**	**100.0**	**100.0**
% Polymeric additive A (solid on pigment)	1.5	20.0	5.0	5.0	20.0	5.0	7.0	20.0	25.2	20.0	30.0
% Polymeric additive A (delivery form on pigment)	5.1	66.6	16.7	16.7	66.8	16.6	23.4	66.5	84.0	66.5	100.0

** Non-volatile matter.
* Synthacryl SC 370 (75%)

Working method: Premix ingredients 1–3, then add aerosil and pigment. Pigmentation level depends on pigment selection. In case of rheologically unfavourable pigments, usage levels may be lower than stated here. Please note that, for simplification purposes, the ingredients in this table (and in related tables) are not numbered; however, "Ingredient #1" is always the topmost ingredient. Other ingredients would be consecutively numbered from top to bottom in each respective table.

Table I.2 High Quality Industrial Coatings: Formulation No. 2, Starting Formulations for Pigment Concentrates with Acrylic Resin and Polymeric Additive C

	Titanium dioxide PW 6	Organic yellow PY 151	Iron oxide yellow PY 42	Chrome yellow PY 34	Organic red PR 170	Molybdate orange PR 104	Iron oxide red PR 101	Phthalo-cyanine green PG 7	Phthalo-cyanine blue PB 15	Carbon black PBk 7
Acrylic resin*	23.0	30.0	22.0	23.0	22.0	21.0	16.5	30.0	30.0	32.0
Methoxypropylacetate	5.8	19.4	6.6	5.0	12.7	0.0	0.0	28.3	36.7	15.1
Polymeric additive C** (45%)	3.0	15.6	9.0	8.4	23.3	14.5	15.1	16.7	13.3	24.9
Aerosil 200	0.2	0.0	0.4	0.6	0.0	0.5	0.4	0.0	0.0	0.0
Pigment	68.0	35.0	62.0	63.0	42.0	64.0	68.0	25.0	20.0	28.0
	100.0	**100.0**	**100.0**	**100.0**	**100.0**	**100.0**	**100.0**	**100.0**	**100.0**	**100.0**
% Polymeric additive C (solid on pigment)	2.0	20.1	6.5	6.0	25.0	10.2	10.0	30.1	29.9	40.0
% Polymeric additive C (delivery form on pigment)	4.4	44.6	14.5	13.3	55.5	22.7	22.2	66.8	66.5	88.9

* Acryloid DM 55 (60%), Synthacryl SC 370 (75%), Acryloid AT-410.

** Non-volatile matter.

Working method: Premix ingredients 1–3, then add aerosil and pigment. Pigmentation level depends on pigment selection. In case of rheologically unfavourable pigments, usage levels may be lower than stated here. Please note that, for simplification purposes, the ingredients in this table (and in related tables) are not numbered; however, "Ingredient #1" is always the topmost ingredient. Other ingredients would be consecutively numbered from top to bottom in each respective table.

Table I.3 High Quality Industrial Coatings: Formulation No. 3. Starting Formulations for Pigment Concentrates with Acrylic Resin and Polymeric Additive E

	Titanium dioxide PW 6	Organic yellow PY 151	Iron oxide yellow PY 42	Chrome yellow PY 34	Organic red PR 170	Molybdate orange PR 104	Iron oxide red PR 101	Organic red–violet PV 23	Phthalo-cyanine green PG 36	Phthalo-cyanine blue PB 15	Carbon black PBk 7
Acrylic resin*	26.0	30.0	27.0	24.8	25.0	23.7	19.3	45.0	30.0	45.0	30.0
Methoxypropylacetate	6.4	18.8	7.7	5.7	16.4	3.3	4.6	25.7	30.0	25.7	22.4
Polymeric additive E** (43%)	2.3	16.2	6.8	5.0	18.6	7.5	10.6	9.3	14.7	9.3	19.6
Aerosil 200	0.3	0.0	0.5	0.5	0.0	0.5	0.5	0.0	0.0	0.0	0.0
Pigment	65.0	35.0	58.0	64.0	40.0	65.0	65.0	20.0	25.0	20.0	28.0
	100.0	**100.0**	**100.0**	**100.0**	**100.0**	**100.0**	**100.0**	**100.0**	**100.0**	**100.0**	**100.0**
% Polymeric additive E (solid on pigment)	1.5	19.9	5.0	3.4	20.0	5.0	7.0	20.0	25.3	20.0	30.1
% Polymeric additive E (delivery form on pigment)	3.5	46.3	11.7	7.8	46.5	11.5	16.3	46.5	58.8	46.5	70.0

* Synthacryl SC 370 (75%).
** Non-volatile matter.

Working method: Premix ingredients 1–3, then add aerosil and pigment. Pigmentation level depends on pigment. Pigmentation level depends on pigment selection. In case of rheologically unfavourable pigments, usage levels may be lower than stated here. Please note that, for simplification purposes, the ingredients in this table (and in related tables) are not numbered; however, "Ingredient #1" is always the topmost ingredient. Other ingredients would be consecutively numbered from top to bottom in each respective table.

Table I.4 High Quality Industrial Coatings: Formulation No. 4. Starting Formulations for Pigment Concentrates with Acrylic Resin and Polymeric Additive H

	Titanium dioxide PW 6	Iron oxide yellow PY 42	Chrome yellow PY 151	Iron oxide red PR 101	Organic red PR 170	Organic red–violet PV 19	Phthalo-cyanine green PG 7	Phthalo-cyanine blue PB 15	Carbon black PBk 7
Acrylic resin*	24.0	21.0	29.0	17.0	25.0	35.0	35.0	30.0	30.0
Methoxypropylacetate	8.1	3.2	16.4	1.2	9.3	30.6	20.6	34.0	14.4
Polymeric additive H** (52%)	2.6	10.3	9.3	11.3	15.4	9.2	11.1	10.9	15.3
Aerosil 200	0.3	0.5	0.3	0.5	0.3	0.2	0.3	0.1	0.3
Blanc fix micro	0.0	15.0	15.0	15.0	20.0	10.0	15.0	7.0	15.0
Pigment	65.0	50.0	30.0	55.0	30.0	15.0	18.0	18.0	25.0
	100.0	**100.0**	**100.0**	**100.0**	**100.0**	**100.0**	**100.0**	**100.0**	**100.0**
% Polymeric additive H (solid on pigment)	2.0	10.7	16.0	10.7	26.7	31.9	32.1	31.5	31.8
% Polymeric additive H (delivery form on pigment)	4.0	20.6	31.0	20.6	51.3	61.3	61.7	60.6	61.2

* Acryloid DM 55 (60%), Synthacryl SC 370 (75%), Acryloid AT-410.

** Non-volatile matter.

Working method: Premix ingredients 1–3, then add aerosil and pigment. Pigmentation level depends on pigment selection. In case of rheologically unfavourable pigments, usage levels may be lower than stated here. Please note that, for simplification purposes, the ingredients in this table (and in related tables) are not numbered; however, "Ingredient #1" is always the topmost ingredient. Other ingredients would be consecutively numbered from top to bottom in each respective table.

Table I.5 General Industrial Coatings: Formulation No. 5. Starting Formulations for Pigment Concentrates with Acrylic Resin and Polymeric Additives E and K

	Titanium dioxide PW 6	Organic yellow PY 151	Iron oxide yellow PY 42	Chrome yellow PY 34	Organic red PR 170	Molybdate orange PR 104	Iron oxide red PR 101	Phthalo-cyanine green PG 7	Phthalo-cyanine blue PB 15	Carbon black PBk 7
Acrylic resin*	23.0	30.0	22.0	23.0	22.0	21.0	16.5	30.0	30.0	32.0
Methoxypropylacetate	5.8	19.4	6.6	5.0	12.7	0.0	0.0	28.3	36.7	15.1
Polymeric additive E** (43%)	2.2	11.7	6.7	6.3	17.5	10.9	11.3	12.5	10.0	18.7
Polymeric additive K** (88%)	0.8	3.9	2.3	2.1	5.8	3.6	3.8	4.2	3.3	6.2
Aerosil 200	0.2	0.0	0.4	0.6	0.0	0.5	0.4	0.0	0.0	0.0
Pigment	68.0	35.0	62.0	63.0	42.0	64.0	68.0	25.0	20.0	28.0
	100.0	**100.0**	**100.0**	**100.0**	**100.0**	**100.0**	**100.0**	**100.0**	**100.0**	**100.0**
% Polymeric additive E (solid on pigment)	1.4	14.4	4.7	4.3	18.0	7.3	7.2	21.5	21.5	28.7
% Polymeric additive K (solid on pigment)	1.0	9.8	3.3	2.9	12.2	5.0	4.9	14.8	14.5	19.5
% Polymeric additive E/K ([3 : 1] delivery form on pigment)	4.4	44.6	14.5	13.3	55.5	22.7	22.2	66.8	66.5	88.9

* Acryloid DM 55 (60%), Synthacryl SC 370 (75%), Acryloid AT-410.

** Non-volatile matter.

Working method: Premix ingredients 1–4, then add aerosil and pigment. Pigmentation level depends on pigment selection. In case of rheologically unfavourable pigments, usage levels may be lower than stated here. Please note that, for simplification purposes, the ingredients in this table (and in related tables) are not numbered; however, "Ingredient #1" is always the topmost ingredient. Other ingredients would be consecutively numbered from top to bottom in each respective table.

Table I.6 General Industrial Coatings: Formulation No. 6. Starting Formulations for Pigment Concentrates with Aldehyde Resin and Polymeric Additive D

	Titanium dioxide PW 6	Organic yellow PY 151	Iron oxide yellow PY 42	Chrome yellow PY 34	Organic red PR 170	Molybdate orange PR 104	Iron oxide red PR 101	Organic red–violet PR 88	Phthalocyanine green PG 7	Phthalocyanine blue PB 15	Carbon black PBk 7
Acrylic resin*	23.0	26.0	20.0	24.0	21.0	25.0	19.0	25.0	25.0	32.0	26.0
Exxsol D 30	9.5	19.4	14.5	7.2	25.7	4.2	6.8	21.7	20.0	24.7	17.5
Methoxypropylacetate	0.0	0.0	0.0	0.0	0.0	0.0	0.0	0.0	19.6	14.0	14.5
Polymeric additive D** (60%)	2.2	12.6	5.0	6.2	13.3	8.3	8.7	13.3	10.4	7.3	14.0
Aerosil 200	0.3	0.0	0.5	0.6	0.0	0.5	0.5	0.0	0.0	0.0	0.0
Pigment	65.0	42.0	60.0	62.0	40.0	62.0	65.0	40.0	25.0	22.0	28.0
	100.0	**100.0**	**100.0**	**100.0**	**100.0**	**100.0**	**100.0**	**100.0**	**100.0**	**100.0**	**100.0**
% Polymeric additive D (solid on pigment)	2.0	18.0	5.0	6.0	20.0	8.0	8.0	20.0	25.0	19.9	30.0
% Polymeric additive D (delivery form on pigment)	3.4	30.0	8.3	10.0	33.3	13.4	13.4	33.3	41.6	33.2	50.0

* Laropal A 81 (65%), Kunstharz EP-TC (65%).

** Non-volatile matter.

Working method: Premix ingredients 1–4, then add aerosil and pigment. Pigmentation level depends on pigment selection. In case of rheologically unfavourable pigments, usage levels may be lower than stated here. Please note that, for simplification purposes, the ingredients in this table (and in related tables) are not numbered; however, "Ingredient #1" is always the topmost ingredient. Other ingredients would be consecutively numbered from top to bottom in each respective table.

Table I.7 General Industrial Coatings: Formulation No. 7. Starting Formulations for Pigment Concentrates with Aldehyde Resin and Polymeric Additive C

	Titanium dioxide PW 6	Organic yellow PY151	Iron oxide yellow PY 42	Chrome yellow PW 34	Organic red PR 170	Molybdate orange PR 104	Iron oxide red PR 101	Organic red–violet PV 29	Organic violet PV 23	Organic red–violet PV 29	Phthalocyanine green PG 7	Phthalocyanine blue PB 15	Carbon black PBk 7
Aldehyde resin*	22.0	28.0	22.0	25.0	26.0	25.0	25.0	38.0	50.0	37.0	33.0	37.0	40.0
Methoxy-propylacetate	9.7	11.3	9.6	1.9	11.8	1.5	1.2	22.9	20.0	29.7	28.1	31.9	13.3
Polymeric additive C** (45%)	3.0	18.7	8.0	8.5	22.2	11.0	13.3	17.1	12.0	13.3	13.9	11.1	18.7
Aerosil 200	0.3	0.0	0.4	0.6	0.0	0.5	0.5	0.0	0.0	0.0	0.0	0.0	0.0
Pigment	65.0	42.0	60.0	64.0	40.0	62.0	60.0	22.0	18.0	20.0	25.0	20.0	28.0
	100.0	**100.0**	**100.0**	**100.0**	**100.0**	**100.0**	**100.0**	**100.0**	**100.0**	**100.0**	**100.0**	**100.0**	**100.0**
% Polymeric additive C (solid on pigment)	2.1	20.0	6.0	6.0	25.0	8.0	10.0	35.0	30.0	29.9	25.0	25.0	30.1
% Polymeric additive C (delivery form on pigment)	4.6	44.5	13.3	13.3	55.5	17.7	22.2	77.7	66.7	66.5	55.6	55.5	66.8

* Laropal A 81 (65%), Kunstharz EP–TC (65%).

** Non-volatile matter

Working method: Premix ingredients 1–3, then add aerosil and pigment. Pigmentational level depends on the pigment selection. In case of rheologically unfavourable pigments, usage levels may be lower than stated here. Please note that, for simplification purposes, the ingredients in this table (and in related tables) are not numbered; however, "Ingredient # 1" is always the topmost ingredient. Other ingredients would be consecutively numbered from top to bottom in each respective table.

Table I.8 General Industrial Coatings: Formulation No. 8. Starting Formulations for Pigment Concentrates with Aldehyde Resin and Polymeric Additive H

	Titanium dioxide PW 6	Organic yellow PY151	Organic yellow PY 74	Chrome yellow PW 34	Iron oxide yellow PY 42	Organic red PR 170	Iron oxide red PR 101	Organic red PV 19	Organic violet PV 23	Phthalo-cyanine green PG 7	Phthalo-cyanine blue PB 15	Carbon black PBk 7
Aldehyde resin*	23.0	28.0	50.0	26.0	22.0	26.0	25.0	39.0	57.0	40.0	40.0	40.0
Methoxy-propylacetate	4.5	8.3	8.1	2.1	4.2	5.9	1.5	5.5	6.5	12.7	11.9	7.9
Xylene	4.5	8.3	8.1	2.1	4.2	5.9	1.5	5.5	6.4	12.6	11.8	7.9
Polymeric additive H** (52%)	2.5	15.4	8.8	7.2	9.2	20.2	11.5	20.0	12.1	12.7	13.3	16.2
Aerosil 200	0.5	0.0	0.0	0.6	0.4	0.0	0.5	0.0	0.0	0.0	0.0	0.0
Pigment	65.0	40.0	25.0	62.0	60.0	42.0	60.0	30.0	18.0	22.0	23.0	28.0
	100.0	**100.0**	**100.0**	**100.0**	**100.0**	**100.0**	**100.0**	**100.0**	**100.0**	**100.0**	**100.0**	**100.0**
% Polymeric additive H (solid on pigment)	2.0	20.0	18.3	6.0	8.0	25.0	10.0	34.7	35.0	30.0	30.1	30.1
% Polymeric additive H (delivery form on pigment)	3.9	38.5	35.2	11.6	15.3	48.1	19.2	66.7	67.2	57.1	57.8	57.8

* Laropal A 81 (65%), Kunstharz EP–TC (65%)

**Non-volatile matter.

Working method: Premix ingredients 1–3, then add aerosil and pigment. Pigmentational level depends on the pigment selection. In case of rheologically unfavourable pigments, usage levels may be lower than stated here. Please note that, for simplification purposes, the ingredients in this table (and in related tables) are not numbered; however, "Ingredient # 1" is always the topmost ingredient. Other ingredients would be consecutively numbered from top to bottom in each respective table.

Table I.9 General Industrial Coatings: Formulation No. 9. Starting Formulations for Pigment Concentrates with Aldehyde Resin and Polymeric Additives E and K

	Titanium dioxide PW 6	Organic yellow PY151	Iron oxide yellow PY 42	Chrome yellow PW 34	Organic red PR 170	Molybdate orange PR 104	Iron oxide red PR 101	Phthalocyanine green PG 7	Phthalocyanine blue PB 15	Carbon black PBk 7
Aldehyde resin*	20.6	21.3	18.4	21.2	16.0	18.5	16.9	28.0	30.0	23.5
Methoxypropylacetate	5.5	11.3	5.7	1.9	12.5	0.5	3.1	12.1	19.0	12.8
Xylene	5.5	11.2	5.6	1.9	12.5	0.4	3.1	12.0	18.9	12.8
Polymeric additive E** (43%)	2.3	12.2	7.4	7.0	14.0	11.3	8.5	13.0	8.7	15.7
Polymeric additive K** (88%)	0.8	4.0	2.4	2.4	5.0	3.8	2.9	4.4	2.9	5.2
Dispersant P	0.0	0.0	0.0	0.0	0.0	0.0	0.0	0.5	0.5	0.0
Aerosil 200	0.3	0.0	0.5	0.6	0.0	0.5	0.5	0.0	0.0	0.0
Pigment	65.0	40.0	60.0	65.0	40.0	65.0	65.0	30.0	20.0	30.0
	100.0	**100.0**	**100.0**	**100.0**	**100.0**	**100.0**	**100.0**	**100.0**	**100.0**	**100.0**
% Polymeric additive E (solid on pigment)	1.5	13.1	5.3	4.5	15.0	7.5	5.6	18.6	18.7	22.5
% Polymeric additive K (solid on pigment)	1.1	8.8	3.5	3.3	11.0	5.1	3.9	12.9	12.8	15.6
% Polymeric additive E/K ([3 : 1] delivery form on pigment)	4.8	40.5	16.3	14.5	47.5	23.2	17.5	58.0	58.0	70.0

* Laropal A 81 (65%).

** Non-volatile matter.

Working method: Premix ingredients 1–6, then add aerosil and pigment. Pigmentational level depends on the pigment selection. In case of rheologically unfavourable pigments, usage levels may be lower than stated here. Please note that, for simplification purposes, the ingredients in this table (and in related tables) are not numbered; however, "Ingredient # 1" is always the topmost ingredient. Other ingredients would be consecutively numbered from top to bottom in each respective table.

Table I.10 General Industrial Coatings: Formulation No. 10. Starting Formulations for Pigment Concentrates with Aldehyde Resin and Polymeric Additive E

	Titanium dioxide PW 6	Organic yellow PY151	Iron oxide yellow PY 42	Chrome yellow PW 34	Organic red PR 170	Molybdate orange PR104	Iron oxide red PR 101	Organic red violet PV 88	Phthalocyanine green PG 7	Phthalocyanine blue PB 15	Carbon black PBk 7
Aldehyde resin*	21.5	24.0	19.0	21.2	18.0	18.5	17.7	20.5	18.5	30.0	23.5
Methoxypropylacetate	7.2	12.7	8.5	1.6	12.0	1.5	1.2	18.8	34.1	36.0	26.5
Polymeric additive E** (43%)	3.0	18.3	10.0	9.1	25.0	14.5	12.7	15.7	17.4	14.0	20.0
Aerosil 200	0.3	0.0	0.5	0.6	0.0	0.5	0.4	0.0	0.0	0.0	0.0
Pigment	68.0	45.0	62.0	67.5	45.0	65.0	68.0	45.0	30.0	20.0	30.0
	100.0	**100.0**	**100.0**	**100.0**	**100.0**	**100.0**	**100.0**	**100.0**	**100.0**	**100.0**	**100.0**
% Polymeric additive E (solid on pigment)	1.9	17.5	6.9	5.8	23.9	9.6	8.0	15.0	24.9	30.1	28.7
% Polymeric additive E (delivery form on pigment)	4.4	40.7	16.1	13.5	55.6	22.3	18.7	34.9	58.0	70.0	66.7

* Laropal A 81 (65%).

** Non-volatile matter.

Working method: Premix ingredients 1–3, then add aerosil and pigment. Pigmentational level depends on the pigment selection. In case of rheologically unfavourable pigments, usage levels may be lower than stated here. Please note that, for simplification purposes, the ingredients in this table (and in related tables) are not numbered; however, "Ingredient # 1" is always the topmost ingredient. Other ingredients would be consecutively numbered from top to bottom in each respective table.

Table I.11 General Industrial Coatings: Formulation No. 11. Starting Formulations for Pigment Concentrates with Alkyd Resin and Polymeric Additives K and B

	Titanium dioxide PW 6	Iron oxide yellow PY 42	Red–violet PR 19	Phthalocyanine blue PB 15	Carbon black PBk 7	Carbon black FW 200 PBk 7
Alkyd resin (70%)*	22.0	32.0	60.0	48.0	48.0	52.0
Methoxypropylacetate	3.4	4.8	8.8	13.8	10.0	12.3
Xylene	5.0	10.0	13.0	20.0	16.8	21.0
Polymeric additive K (88%)**	0.8	2.0	2.0	2.0	1.5	1.5
Polymeric additive B** (38%)	0.2	0.7	1.0	1.0	3.5	5.0
Defoamer	0.1	0.1	0.1	0.1	0.1	0.1
Anti-gelling agent	0.1	0.1	0.1	0.1	0.1	0.1
Aerosil 200	0.4	0.3	0.0	0.0	0.0	0.0
Pigment	68.0	50.0	15.0	15.0	20.0	8.0
	100.0	**100.0**	**100.0**	**100.0**	**100.0**	**100.0**
% Polymeric additive K/B (solid on pigment)	1.1/0.1	3.5/0.5	11.7/2.5	11.7/2.5	6.6/6.7	16.5/23.8
% Polymeric additive K/B (delivery form on pigment)	1.2/0.3	4.0/1.4	13.3/6.7	13.3/6.7	7.5/17.5	18.8/62.5

* Italkid F 50 P (70%).

** Non-volatile matter.

Working method: Premix ingredients 1–7, then add aerosil and pigment. Pigmentational level depends on the pigment selection. In case of rheologically unfavourable pigments, usage levels may be lower than stated here. Please note that, for simplification purposes, the ingredients in this table (and in related tables) are not numbered; however, "Ingredient # 1" is always the topmost ingredient. Other ingredients would be consecutively numbered from top to bottom in each respective table.

Table I.12 General Industrial Coatings: Formulation No. 12. Starting Formulations for Pigment Concentrates with Alkyd Resin and Polymeric Additive H

	Kronos 2310 PW 6	Flammr. 101 PBk7	FW 200 PBk 7	Heliogen blue L6975 F PB 15	Hostap. red–violet ER 02 PR 19	Bayferrox 3920 PY 42	Special black 4 PBk 7	Hostap. yellow H4G PY 151	Hostap. violet RL–NF PV 23	Sunfast green 264–8735 PG 7
Alkyd Resin*	22.0	48.0	52.0	48.0	60.0	32.0	48.0	50.0	65.0	60.0
Xylene	3.6	16.3	20.9	19.1	12.1	9.2	16.3	9.8	9.0	12.9
Methoxypropylacetate	2.3	10.3	12.7	13.5	8.5	4.4	10.3	8.0	8.5	9.0
Polymeric additive H** (52%)	1.5	5.2	6.2	4.2	4.2	3.9	5.2	2.0	2.3	2.9
Anti-gelling agent	0.1	0.1	0.1	0.1	0.1	0.1	0.1	0.1	0.1	0.1
Defoamer	0.1	0.1	0.1	0.1	0.1	0.1	0.1	0.1	0.1	0.1
Aerosil 200	0.4	0.0	0.0	0.0	0.0	0.3	0.0	0.0	0.0	0.0
Pigment	70.0	20.0	8.0	15.0	15.0	50.0	20.0	30.0	15.0	15.0
	100.0	**100.0**	**100.0**	**100.0**	**100.0**	**100.0**	**100.0**	**100.0**	**100.0**	**100.0**
% Polymeric additive H (solid on pigment)	1.1	13.5	40.3	14.6	14.6	4.1	13.5	3.5	8.0	10.0
% Polymeric additive H (delivery form on pigment)	2.1	26.0	77.5	28.0	28.0	7.8	26.0	6.7	15.3	19.3

* Italkid F 50 P.

** Non-volatile matter.

Working method: Premix ingredients 1–6, then add aerosil and pigment. Pigmentational level depends on the pigment selection. In case of rheologically unfavourable pigments, usage levels may be lower than stated here. Please note that, for simplification purposes, the ingredients in this table (and in related tables) are not numbered; however, "Ingredient # 1" is always the topmost ingredient. Other ingredients would be consecutively numbered from top to bottom in each respective table.

Table I.13 Architectural Coatings: Formulation No. 13. Starting Formulations for Pigment Concentrates with Aldehyde Resin and Polymeric Additive K

	Titanium dioxide PW 6	Iron oxide yellow PY 42	Chrome yellow PY 34	Organic red PR 48	Organic red PR 146	Molybdate orange PR 104	Iron oxide red PR 101	Organic red–violet PV 19	Organic red–violet PV 23	Phthalo-cyanine green PG 7	Chrome oxide green PG 17	Phthalo-cyanine blue PB 15	Carbon black PBk 7
Aldehyde resin*	23.0	20.0	22.0	35.0	38.0	24.0	26.0	37.0	65.0	45.0	32.0	43.0	49.0
Methoxy-propylacetate	7.0	11.2	10.0	22.8	22.3	8.7	7.3	21.3	11.5	17.9	4.5	19.0	18.2
Xylene	7.0	11.2	10.0	22.8	22.3	8.7	7.3	21.2	11.5	17.9	4.5	19.0	18.1
Polymeric additive K** (88%)	0.7	3.1	2.5	1.4	1.4	3.1	3.8	2.5	2.0	1.2	2.5	1.0	1.7
Aerosil 200	0.3	0.5	0.5	0.0	0.0	0.5	0.6	0.0	0.0	0.0	0.5	0.0	0.0
Pigment	62.0	54.0	55.0	18.0	16.0	55.0	55.0	18.0	10.0	18.0	56.0	18.0	13.0
	100.0	**100.0**	**100.0**	**100.0**	**100.0**	**100.0**	**100.0**	**100.0**	**100.0**	**100.0**	**100.0**	**100.0**	**100.0**
% Polymeric additive K (solid on pigment)	1.0	5.1	4.0	6.8	7.7	5.0	6.1	12.2	17.6	5.9	3.9	4.9	11.5
% Polymeric additive K (delivery form on pigment)	1.1	5.7	4.6	7.8	8.8	5.6	6.9	13.9	20.0	6.7	4.5	5.6	13.1

* Laropal A 81 (65%)

**Non-volatile matter.

Working method: Premix ingredients 1–3, then add aerosil and pigment. Pigmentational level depends on the pigment selection. In case of rheologically unfavourable pigments, usage levels may be lower than stated here. Please note that, for simplification purposes, the ingredients in this table (and in related tables) are not numbered; however, "Ingredient # 1" is always the topmost ingredient. Other ingredients would be consecutively numbered from top to bottom in each respective table.

Table I.14 Architectural Coatings: Formulation No. 14. Starting Formulations for Pigment Concentrates with Ketone Resin and Polymeric Additive E and K

	Titanium dioxide PW 6	Organic yellow PY 151	Organic yellow PY 74	Iron oxide yellow PY 42	Organic red–violet PV 19	Molybdate orange PR 104	Iron oxide red PR 101	Chrome oxide green PG 17	Phthalocyanine green PG 7	Phthalocyanine blue PB 15	Carbon black PBk 7
Ketone resin*	30.0	35.0	50.0	30.0	38.0	20.0	30.0	35.0	58.0	48.0	45.0
Exxsol D30	13.0	28.5	13.1	11.5	24.8	17.0	7.3	1.2	16.3	22.4	20.8
Polymeric additive E** (43%)	1.2	3.5	3.5	0.0	3.5	1.7	0.0	0.0	4.0	2.7	9.8
Polymeric additive K** (88%)	0.5	1.0	1.4	3.0	1.7	1.3	4.2	3.3	1.7	1.9	2.4
Aerosil 200	0.3	0.0	0.0	0.5	0.0	0.0	0.5	0.5	0.0	0.0	0.0
Pigment	55.0	32.0	32.0	55.0	32.0	60.0	58.0	60.0	20.0	25.0	22.0
	100.0	**100.0**	**100.0**	**100.0**	**100.0**	**100.0**	**100.0**	**100.0**	**100.0**	**100.0**	**100.0**
% Polymeric additive E (solid on pigment)	0.9	4.7	4.7	0.0	4.7	1.2	0.0	0.0	8.6	4.6	19.2
Polymeric additive K (solid on pigment)	0.8	2.8	3.9	4.8	4.7	1.9	6.4	4.8	7.5	6.7	9.6
% Polymeric additive E/ K ([3:1] delivery form on pigment)	3.1	14.1	15.3	5.5	16.3	5.0	7.2	5.5	28.5	18.4	55.5

* Laropal A 81 (65%).

** Non-volatile matter.

Working method: Premix ingredients 1–4, then add aerosil and pigment. Pigmentational level depends on the pigment selection. In case of rheologically unfavourable pigments, usage levels may be lower than stated here. Please note that, for simplification purposes, the ingredients in this table (and in related tables) are not numbered; however, "Ingredient # 1" is always the topmost ingredient. Other ingredients would be consecutively numbered from top to bottom in each respective table.

Table I.15 Architectural Coatings: Formulation No. 15. Starting Formulations for Pigment Concentrates with Long Oil Alkyd and Polymeric Additive K

	Titanium dioxide PW 6	Organic yellow PY 74	Iron oxide yellow PY 42	Organic red PR 112	Iron oxide red PR 101	Phthalocyanine blue PB 15	Phthalocyanine green PG 7	Carbon black PBk 7
Long oil alkyd*	25.0	40.0	35.0	30.0	30.0	45.0	45.0	45.0
Exxsol D30	13.2	23.0	11.9	37.5	5.7	24.6	27.6	24.0
Polymeric additive K** (88%)	1.0	1.5	2.2	2.0	3.3	1.9	1.9	2.5
Ascinin R	0.5	0.5	0.5	0.5	0.5	0.5	0.5	0.5
Aerosil 200	0.3	0.0	0.4	0.0	0.5	0.0	0.0	0.0
Pigment	60.0	35.0	50.0	30.0	60.0	28.0	25.0	28.0
	100.0	**100.0**	**100.0**	**100.0**	**100.0**	**100.0**	**100.0**	**100.0**
% Polymeric additive K (solid on pigment)	1.5	3.8	3.9	5.9	4.8	6.0	6.7	7.9
Polymeric additive K (delivery form on pigment)	1.7	4.3	4.4	6.7	5.5	6.8	7.6	8.9

* Worléekyd T 768.

** Non-volatile matter.

Working method: Premix ingredients 1–4, then add aerosil and pigment. Pigmentational level depends on the pigment selection. In case of rheologically unfavourable pigments, usage levels may be lower than stated here. Please note that, for simplification purposes, the ingredients in this table (and in related tables) are not numbered; however, "Ingredient # 1" is always the topmost ingredient. Other ingredients would be consecutively numbered from top to bottom in each respective table.

Table I.16 Architectural Coatings: Formulation No. 16. Starting Formulations for Pigment Concentrates with Short Oil Alkyd and Polymeric Additive C

	Titanium dioxide PW 6	Organic yellow PY 151	Iron oxide yellow PY 42	Chrome yellow PY 34	Organic red PR 170	Molybdate orange PR 104	Iron oxide red PR 101	Organic red–violet PV 29	Phthalo-cyanine green PG 36	Phthalo-cyanine blue PB 15	Carbon black PBk 7
Short oil alkyd*	21.5	24.0	19.2	21.2	18.0	21.0	16.1	25.7	26.0	27.5	23.5
Shellsol A	8.1	13.5	11.5	2.0	17.0	0.0	1.0	31.0	38.3	36.5	23.5
Polymeric additive C** (45%)	2.2	17.5	7.0	8.7	20.0	11.5	12.5	13.3	12.7	11.0	19.0
Aerosil 200	0.2	0.0	0.3	0.6	0.0	0.5	0.4	0.0	0.0	0.0	0.0
Pigment	68.0	45.0	62.0	67.5	45.0	67.0	70.0	30.0	23.0	25.0	34.0
	100.0	**100.0**	**100.0**	**100.0**	**100.0**	**100.0**	**100.0**	**100.0**	**100.0**	**100.0**	**100.0**
% Polymeric additive C (solid on pigment)	1.5	17.5	5.1	5.8	20.0	7.7	8.0	20.0	24.9	19.8	25.2
% Polymeric additive C (delivery form on pigment)	3.2	38.9	11.3	12.9	44.4	17.2	17.9	44.3	55.2	44.0	55.9

* Alfatalat A M 318 (70%).
**Non-volatile matter.
Working method: Premix ingredients 1–3, then add aerosil and pigment. Pigmentational level depends on the pigment selection. In case of rheologically unfavourable pigments, usage levels may be lower than stated here. Please note that, for simplification purposes, the ingredients in this table (and in related tables) are not numbered; however, "Ingredient # 1" is always the topmost ingredient. Other ingredients would be consecutively numbered from top to bottom in each respective table.

Table I.17 Wood Coatings: Formulation No. 17. Starting Formulations for Pigment Concentrates with Short Oil Alkyd and Polymeric Additive K and E

	Kronos 2310 PW 6	Flammr. 101 PBk 7	FW 200 PBk 7	Heliogen blue L 6975F PB 15	Hostaperm red-violet ER 02 PV 19	Bayferrox 3920 PY 42	Special black 4 PBk 7
Short oil alkyd*	25.0	45.0	50.5	45.0	56.0	30.0	45.0
Xylene	5.1	19.8	21.4	22.0	16.0	11.6	19.8
Methoxypropylacetate	3.5	10.0	12.3	14.8	9.8	5.2	10.0
Polymeric additive K** (88%)	0.8	1.5	1.8	2.0	2.0	2.0	1.5
Polymeric additive E** (43%)	0.2	3.5	5.0	1.0	1.0	0.7	3.5
Anti-gelling agent	0.1	0.1	0.1	0.1	0.1	0.1	0.1
Defoamer	0.1	0.1	0.1	0.1	0.1	0.1	0.1
Aerosil 200	0.2	0.0	0.0	0.0	0.0	0.3	0.0
Pigment	65.0	20.0	8.8	15.0	15.0	50.0	20.0
	100.0	**100.0**	**100.0**	**100.0**	**100.0**	**100.0**	**100.0**
% Polymeric additive K (solid on pigment)	1.1	6.6	18.0	11.7	11.7	3.5	6.6
% Polymeric additive E (solid on pigment)	0.1	7.5	24.4	2.9	2.9	0.6	7.5
% Polymeric additive K/E (delivery form on pigment)	1.5	25.0	77.3	20.0	20.0	5.4	25.0

* Italkid F 29.

** Non-volatile matter.

Working method: Premix ingredients 1–7, then add aerosil and pigment. Pigmentational level depends on the pigment selection. In case of rheologically unfavourable pigments, usage levels may be lower than stated here. Please note that, for simplification purposes, the ingredients in this table (and in related tables) are not numbered; however, "Ingredient # 1" is always the topmost ingredient. Other ingredients would be consecutively numbered from top to bottom in each respective table.

Table I.18 Wood Coatings: Formulation No. 18. Starting Formulations for Pigment Concentrates with Short Oil Alkyd and Polymeric Additive H

	Titanium dioxide PW 6	Organic yellow PY 151	Iron oxide yellow PY 42	Organic red–violet PV 19	Organic violet PV 23	Phthalo-cyanine green PG 7	Phthalo-cyanine blue PB 15	Carbon black PBk 7	Carbon black FW 200 PBk 7
Short oil alkyd*	25.0	45.0	30.0	56.0	62.0	57.0	45.0	45.0	50.5
Methoxypropylacetate	3.4	9.0	4.9	9.5	9.0	9.0	14.5	10.0	12.8
Xylene	4.7	13.8	10.7	15.1	11.5	15.9	21.1	19.3	20.9
Polymeric additive H** (52%)	1.5	2.0	3.9	4.2	2.3	2.9	4.2	5.2	6.8
Anti-gelling agent	0.1	0.1	0.1	0.1	0.1	0.1	0.1	0.1	0.1
Defoamer	0.1	0.1	0.1	0.1	0.1	0.1	0.1	0.1	0.1
Aerosil 200	0.2	0.0	0.3	0.0	0.0	0.0	0.0	0.0	0.0
Pigment	65.0	30.0	50.0	15.0	15.0	15.0	15.0	20.0	8.8
	100.0	**100.0**	**100.0**	**100.0**	**100.0**	**100.0**	**100.0**	**100.0**	**100.0**
% Polymeric additive H (solid on pigment)	1.2	3.5	4.1	14.6	8.0	10.1	14.6	13.5	40.2
% Polymeric additive H (delivery form on pigment)	2.3	6.7	7.8	28.0	15.3	19.3	28.0	26.0	77.3

* Italkid F 29 (75%).

** Non-volatile matter.

Working method: Premix ingredients 1–6, then add aerosil and pigment. Pigmentational level depends on the pigment selection. In case of rheologically unfavourable pigments, usage levels may be lower than stated here. Please note that, for simplification purposes, the ingredients in this table (and in related tables) are not numbered; however, "Ingredient # 1" is always the topmost ingredient. Other ingredients would be consecutively numbered from top to bottom in each respective table.

Table I.19 Wood Coatings: Formulation No. 19. Starting Formulations for Pigment Concentrates with Polyester Resin and Polymeric Additive A

	Titanium dioxide PW 6	Organic yellow PY 151	Iron oxide yellow PY 42	Chrome yellow PY 34	Organic red PR 170	Molybdate orange PR 104	Iron oxide red PR 101	Organic red–violet PR 122	Phthalo-cyanine green PG 36	Phthalo-cyanine blue PB 15	Carbon black PBk 7
Polyester resin*	22.5	40.0	22.0	22.5	26.4	19.0	27.0	34.7	30.0	35.0	25.0
Shellsol A	4.5	7.5	7.4	3.9	10.8	3.5	1.5	8.0	25.4	24.2	7.5
Methoxypropylacetate	4.5	7.5	6.0	3.2	4.5	2.0	1.2	24.0	8.0	7.5	17.5
Polymeric additive A** (30%)	3.2	15.0	9.2	10.0	23.3	15.0	10.0	13.3	16.6	13.3	25.0
Aerosil 200	0.3	0.0	0.4	0.4	0.0	0.5	0.3	0.0	0.0	0.0	0.0
Pigment	65.0	30.0	55.0	60.0	35.0	60.0	60.0	20.0	20.0	20.0	25.0
	100.0	100.0	100.0	100.0	100.0	100.0	100.0	100.0	100.0	100.0	100.0
% Polymeric additive A (solid on pigment)	1.5	15.0	5.0	5.0	20.0	7.5	5.0	20.0	24.9	20.0	30.0
% Polymeric additive A (delivery form on pigment)	4.9	50.0	16.7	16.7	66.6	25.0	16.7	66.6	83.0	67.0	100.0

* Desmophen RD 181 (75%).

** Non-volatile matter.

Working method: Premix ingredients 1–4, then add aerosil and pigment. Pigmentational level depends on the pigment selection. In case of rheologically unfavourable pigments, usage levels may be lower than stated here. Please note that, for simplification purposes, the ingredients in this table (and in related tables) are not numbered; however, "Ingredient # 1" is always the topmost ingredient. Other ingredients would be consecutively numbered from top to bottom in each respective table.

Table I.20a Aqueous Coatings: Formulation No. 20a. Starting Formulations for Resin-Free Pigment Concentrates with Polymeric Additive G

	Kronos 2160 PW 6	Bayferrox yellow 3920 PY 42	Hostap. yellow H3G PY 154	Brilliant yellow 2 GX70 PY 74	Novoperm orange HL 70 NF PO 36	Bayferrox red 130 FS PR 101	Hostaper mrosa E PR 122	Novop. red F 3 RK 70 PR 122	Irgazin red DPP-BO PR 254
Water	19.8	23.0	37.6	28.9	46.4	23.6	46.4	33.9	38.9
Polymeric additive G* (40%)	8.8	20.6	26.3	30.0	22.5	15.0	22.5	25.0	30.0
Defoamer	1.0	1.0	1.0	1.0	1.0	1.0	1.0	1.0	1.0
Proxel GXL	0.1	0.1	0.1	0.1	0.1	0.1	0.1	0.1	0.1
Aerosil 200	0.3	0.3	—	—	—	0.3	—	—	—
Pigment	70.0	55.0	35.0	40.0	30.0	60.0	30.0	40.0	30.0
	100.0	**100.0**	**100.0**	**100.0**	**100.0**	**100.0**	**100.0**	**100.0**	**100.0**
% Polymeric additive G (solid on pigment)	5.0	15.0	30.1	30.0	30.0	10.0	30.0	25.0	40.0
% Polymeric additive G (delivery form on pigment)	12.5	37.5	75.1	75.0	75.0	25.0	75.0	62.5	100.0

* Non-volatile matter.

Working method: Premix ingredients 1–4, then add aerosil and pigment. Pigmentational level depends on the pigment selection. In case of rheologically unfavourable pigments, usage levels may be lower than stated here. Please note that, for simplification purposes, the ingredients in this table (and in related tables) are not numbered; however, "Ingredient # 1" is always the topmost ingredient. Other ingredients would be consecutively numbered from top to bottom in each respective table.

Table I.20b Aqueous Coatings: Formulation No. 20b. Starting Formulations for Resin-Free Pigment Concentrates with Polymeric Additive G

	Hostaperm Violet RL spez. PV 23	Hostaperm V. ER 02 PV 19	Hostaperm red E3B PV 19	Heliog. Blue L6975 F PB 15:2	Heliog. Green L8730 PG 7	Carbon Black FW 200 PBk 7	Carbon Black FW 285 PBk 7	Monarch 110 PBk 7	Elftex 415 PBk 7	Raven 5000 PBk 7
Water	46.4	46.4	46.4	28.9	33.9	57.6	55.6	38.4	33.9	49.4
Polymeric additive G (40%)*	22.5	22.5	22.5	35.0	25.0	26.3	26.3	38.5	35.0	31.5
Defoamer	1.0	1.0	1.0	1.0	1.0	1.0	1.0	1.0	1.0	1.0
Proxel GXL	0.1	0.1	0.1	0.1	0.1	0.1	0.1	0.1	0.1	0.1
Aerosil 200	—	—	—	—	—	—	—	—	—	—
Pigment	30.0	30.0	30.0	35.0	40.0	15.0	15.0	22.0	20.0	18.0
	100.0	100.0	100.0	100.0	100.0	100.0	100.0	100.0	100.0	100.0
% Polymeric additive G (solid on pigment)	30.0	30.0	30.0	40.0	25.0	70.1	70.1	70.0	70.0	70.0
% Polymeric additive G (delivery form on pigment)	75.0	75.0	75.0	100.0	62.5	175.3	175.3	175.0	175.0	175.0

* Non-volatile matter.

Working method: Premix ingredients 1–4, then add aerosil and pigment. Pigmentational level depends on the pigment selection. In case of rheologically unfavourable pigments, usage levels may be lower than stated here. Please note that, for simplification purposes, the ingredients in this table (and in related tables) are not numbered; however, "Ingredient # 1" is always the topmost ingredient. Other ingredients would be consecutively numbered from top to bottom in each respective table.

Formulation Set #	Type of Formulation
I.9	**General Industrial Coatings/Architectural Coatings** Starting formulations for Pigment Concentrates with **ALDEHYDE RESIN** and Polymeric Additives E & K
I.10	**General Industrial Coatings** Starting formulations for Pigment Concentrates with **ALDEHYDE RESIN** and Polymeric Additive E
I.11	**General Industrial Coatings** Starting formulations for Pigment Concentrates with **ALKYD RESIN** and Polymeric Additives K & B
I.12	**General Industrial Coatings** Starting formulations for Pigment Concentrates with **ALKYD RESIN** and Polymeric Additive H
I.13	**Architectural Coatings** Starting formulations for Pigment Concentrates with **ALDEHYDE RESIN** and Polymeric Additive K
I.14	**Architectural Coatings** Starting formulations for Pigment Concentrates with **KETONE RESIN** and Polymeric Additives E & K
I.15	**Architectural Coatings** Starting formulations for Pigment Concentrates with **ALKYD RESIN** and Polymeric Additive K
I.16	**Architectural Coatings** Starting formulations for Pigment Concentrates with **ALKYD RESIN** and Polymeric Additive C
I.17	**Wood Coatings** Starting formulations for Pigment Concentrates with **ALKYD RESIN** and Polymeric Additives K & E

Figure I.2 Description of the formulation sets (Part 2 of 3)

Formulation Set #	Type of Formulation
I.1	**High Quality Industrial Coatings** Starting formulations for Pigment Concentrates with **ACRYLIC RESIN** and Polymeric Additive A
I.2	**High Quality Industrial Coatings/General Industrial Coatings** Starting formulations for Pigment Concentrates with **ACRYLIC RESIN** and Polymeric Additive C
I.3	**High Quality Industrial Coatings** Starting formulations for Pigment Concentrates with **ACRYLIC RESIN** and Polymeric Additive E
I.4	**High Quality Industrial Coatings** Starting formulations for Pigment Concentrates with **ACRYLIC RESIN** and Polymeric Additive H
I.5	**High Quality Industrial Coatings** Starting formulations for Pigment Concentrates with **ACRYLIC RESIN** and Polymeric Additives E & K
I.6	**General Industrial Coatings/Architectural Coatings** Starting formulations for Pigment Concentrates with **ALDEHYDE RESIN** and Polymeric Additive D
I.7	**General Industrial Coatings** Starting formulations for Pigment Concentrates with **ALDEHYDE RESIN** and Polymeric Additive C
I.8	**General Industrial Coatings/Architectural Coatings** Starting formulations for Pigment Concentrates with **ALDEHYDE RESIN** and Polymeric Additive H

Figure I.1 Description of the formulation sets (Part 1 of 3)

Formulation Set #	Type of Formulation
I.18	**Wood Coatings** Starting formulations for Pigment Concentrates with **ALKYD RESIN** and Polymeric Additive H
I.19	**Wood Coatings** Starting formulations for Pigment Concentrates with **POLYESTER RESIN** and Polymeric Additive A
I.20a	**Aqueous Coatings** Starting formulations for **RESIN-FREE** Pigment Concentrates with Polymeric Additive G
I.20b	**Aqueous Coatings** Starting formulations for **RESIN-FREE** Pigment Concentrates with Polymeric Additive G

Please note that each formulation set is composed of multiple formulation possibilities (often 8-12 individual formulations); a generic description of selected additives is included at the end of Appendix I.

Figure I.3 Description of the formulation sets (Part 3 of 3)

Table I.21 Additive Cross-Reference Charts

Polymeric additive*	Description	Typical physical data						Solvents	Flash point (setaflash) C (F)
		Amine value mg KOH/g	Acid value mg KOH/g	Density (20 C/g ml)	Weight/ U.S. gal. (lb/gal.)	Refractive index	Non-volatile matter		
A	A high molecular-weight copolymer with pigment affinic groups that provide both steric hindrance and electrostatic charge repulsion. Within the family of products represented by products "A", "B", "C" and "D"— additive "A" has both the highest molecular weight and the highest polarity. High quality coatings, including topcoats, benefit from the *triple* features (wetting and dispersing, followed by energy-state stabilization) offered by the aforementioned family of products	11	—	1.02	8.44	1.438	30.0	1-Methoxy-2-propanol acetate/butyl acetate 6/1	38 (100)
B	Has a slightly lower molecular weight and polarity than additive "A"	13	—	1.01	8.36	1.470	38.0	1-Methoxy-2-propanol acetate/xylene/butyl acetate 5/4/2	28 (82)
C	Has an even lower molecular weight than additive "B", is less polar and often demonstrates the broadest compatibilty in pigment concentrates	10	—	0.99	8.26	1.480	45.0	Xylene/butyl/acetate 1-methoxy-2propanol-acetate 3/1/1/	28 (82)
D	Has a molecular weight similar to that of additive "C", but is less polar and its solids content is higher. In addition, it does not contain aromatics	18	—	1.03	8.53	—	60.0	Butyl acetate	26 (79)

* Tables I.21 to I.23 include only selected products requiring additional descriptive information.

Table I.22 Additive Cross-Reference Charts

Polymeric additive*	Description	Typical physical data							
		Amine value mg KOH/g	Acid value mg KOH/g	Density (20 C/g/ml)	Weight/U.S. gal. (lb/gal.)	Refractive index	Non-volatile matter	Solvents	Flash point (setaflash) C (F)
E	Contains not only copolymeric functionalities, but also special pigment affinic groups which demonstrate tailored compatibility with aqueous and solvent-based systems	14	—	1.03	8.57	—	43.0	1-Methoxy-2-propanol acetate/2-methoxy methylethoxy propanol/butyl acetate 7/4/4	38 (100)
G	Utilizes polyfunctional moieties with anionic/non-ionic character. This product contains no organic solvent; it was designed for the production of resin-free, stable pigment concentrates for flood/float-free aqueous systems	10		1.06	8.81	1.400	40.0	Water	>100 (>212)
H	A copolymeric moiety specially designed for general industrial coatings, high quality architectural paints, and pigment concentrate formulations. It has a wide compatibility range and reduces the mill base viscosity	24	—	0.96	7.99	1.480	52.0	Xylene/butyl acetate/1-Methoxy-2-propanol acetate 5/1/1	>24 (>75)
J	Contains a copolymer with acidic groups; this product reduces mill base viscosity and is especially suitable for stabilizing inorganic pigments such as titanium dioxide	—	52	1.03	8.53	1.460	52.0	1-Methoxy-2-propanol acetate/naphtha 1/1	42 (108)

*See footnote to Table I.21.

Table I.23 Additive Cross-Reference Charts

Polymeric additive*	Description	Typical physical data							
		Amine value mg KOH/g	Acid value mg KOH/g	Density (20 C/g/ml)	Weight/ U.S. gal. (lb/gal.)	Refractive index	Non-volatile matter	Solvents	Flash point (setaflash) C (F)
K	Contains a hydroxyfunctional carboxylic acid ester with pigment affinic groups. This product was especially developed to stabilize titanium dioxide, extenders and other inorganic and organic pigments. Often the extender content can be increased without losing gloss. Mill base viscosity is reduced drastically. White base paints stabilized with additive "K" have better colour acceptance and do not demonstrate flooding and floating when tinted with universal colorants.	64	—	0.92	7.65	1.476	88.0	Isoalkanes	54 (129)
L	More polar than additive "E", and is primarily for aqueous systems. Although it demonstrates compatibility with a wide range of pigments, its broadest spectrum of compatibility occurs with organic pigments	14	—	1.09	9.03	1.475	52.0	2-Methoxy-methylethoxy propanol/propylene glycol 2/1	>78 (>172)
M	Because product "M" is a permutation of product "A", please refer to the previously recorded information.								

*See footnote to Table I.21.

Appendix II: Additive Usage Levels for 250 Different Pigments

More than 250 different pigments have been evaluated in extensive laboratory test batteries; the results can be found in Tables II.1 to II.21. (These 21 tables originally appeared as a 114-table set, but they have been condensed into their current form.)

Although a rather comprehensive cross-section of pigments and manufacturers is shown, space limitations prohibit the provision of an exhaustive listing. Several hundred pages would be required for inclusion of the 4000+ commercially available pigments. Through utilization of the many raw material directories available in the industry, one can easily "fill the gap", so to speak, for unlisted pigments merely by cross-referencing the unlisted product with its nearest counterpart in this appendix.

Pigments are listed in alphabetical order together with the name of the supplier and the color index; please note that, for occasional products, two or more widely accepted alternate spellings may exist. Where available, BET surface area (m^2/g) and the recommended pigment/binder ratio are given.

The final columns contain the recommended dosages of several additives (delivery form upon pigment weight). The usage levels are highly dependent upon pigment particle size; optimal levels are determined through a series of laboratory tests.

All information and data stated herein, although in no way guaranteed, are based upon tests and reports considered to be reliable and are believed to be accurate. No warranty, either express or implied, is made or intended. Use should be based upon one's own investigations and appraisals. Any recommendation should not be construed as an invitation to use a material in infringement of patents. Neither the author nor the publisher accepts responsibility for printing errors. Please note that certain information—in the body and/or appendices of this textbook—is based on data reported by outside parties (for instance, by chemical suppliers, distributors, etc.). Neither the inclusion nor the exclusion of various products is indicative of endorsement or non-endorsement by the author or publisher. Consistent with standard reference book publishing conventions, any and all trade names are the respective properties of their owner(s); no entry in this reference work should be regarded as affecting the validity of any trademark or service mark; accordingly, trademark symbols are not used. Please note the author has observed the occurrence of an absolute plethora of different spellings, punctuations, and unit-to-unit conversion rules for certain technical terms and chemical nomenclature expressions in the industry. As paradoxical as it may seem, often even the accepted industry reference books and dictionaries differ greatly on certain conversion rules, spelling issues, "pluralization conventions", abbreviation shortcuts, hyphenation/punctuation rules and word-division protocols. Accordingly, several alternate conversion rules and/or grammatical usage practices—all equally correct—may indeed be employed by some readers. The aforementioned statements apply not only to the appendices, but also to all other materials in this textbook.

Table II.1

Pigment (English)	Pigment (German)	Supplier	Color index	BET	Pigment/ binder	Polymeric additive							
						K	J	H	B	C	D	E	G
Bayertitan R-D	Bayertitan R-D	Bayer	W	6	22	4–5	3	6	5	7	6	4	6	5
Bayertitan R-FD–1	Bayertitan R-FD–1	Bayer	W	6	15	4–5	3	6	5	7	6	4	6	5
Bayertitan R-KB-2	Bayertitan R-KB-2	Bayer	W	6	12	4–5	3	6	5	7	6	4	6	5
Bayertitan R-KB-3	Bayertitan R-KB-3	Bayer	W	6	16	4–5	3	6	5	7	6	4	6	5
Bayertitan R-KB-4	Bayertitan R-KB-4	Bayer	W	6	17	4–5	3	6	5	7	6	4	6	5
Bayertitan R-KB-5	Bayertitan R-KB-5	Bayer	W	6		4–5	3	6	5	7	6	4	6	5
Bayertitan R-KB-6	Bayertitan R-KB-6	Bayer	W	6		4–5	3	6	5	7	6	4	6	5
Bayferrox red 105 M	Bayferrox 105 M	Bayer	R	101		4.5–5	6	19	19	26	22	17	23	38
Bayferrox red 110 M	Bayferrox 110 M	Bayer	R	101		4.5–5	6	19	19	26	22	17	23	38
Bayferrox red 120	Bayferrox 120	Bayer	R	101		4.5–5	6	19	19	26	22	17	23	38
Bayferrox red 120 FS	Bayferrox 120 FS	Bayer	R	101		4.5–5	6	19	19	26	22	17	23	38
Bayferrox red 120 M	Bayferrox 120 M	Bayer	R	101		4.5–5	6	19	19	26	22	17	23	38
Bayferrox red 120 N	Bayferrox 120 N	Bayer	R	101		4.5–5	6	19	19	26	22	17	23	38
Bayferrox red 120 NM	Bayferrox 120 NM	Bayer	R	101	15	4.5–5	6	19	19	26	22	17	23	38
Bayferrox 130	Bayferrox 130	Bayer	R	101		4.5–5	6	19	19	26	22	17	23	38

Table II.2

Pigment (English)	Pigment (German)	Supplier	Color index	BET	Pigment/ binder	Polymeric additive							
						K	J	H	B	C	D	E	G
Bayferrox red 130 BM	Bayferrox 130 BM	Bayer	R 101		4.5–5	9	15	15	21	18	13	19	20
Bayferrox red 130 M	Bayferrox 130 M	Bayer	R 101		4.5–5	6	19	19	26	22	17	23	25
Bayferrox red 140 M	Bayferrox 140 M	Bayer	R 101		4.5–5	6	19	19	26	22	17	23	38
Bayferrox red 160	Bayferrox 160	Bayer	R 101		4.5–5	6	19	19	26	22	17	23	38
Bayferrox red 160 M	Bayferrox 160 M	Bayer	R 101		4.5–5	6	19	19	26	22	17	23	38
Bayferrox red 180	Bayferrox 180	Bayer	R 101		4.5–5	6	19	19	26	22	17	23	38
Bayferrox red 180 M	Bayferrox 180 M	Bayer	R 101		4.5–5	6	19	19	26	22	17	23	38
Bayferrox yellow 3420	Bayferrox 3420	Bayer	Y 42		4.2–5	6	14	14	19	17	13	17	38
Bayferrox yellow 3910	Bayferrox 3910	Bayer	Y 42		4.2–5	6	14	14	19	17	13	17	38
Bayferrox yellow 910	Bayferrox 910	Bayer	Y 42		4.2–5	6	14	14	19	17	13	17	38
Bayferrox yellow 915	Bayferrox 915	Bayer	Y 42		4.2–5	6	14	14	19	17	13	17	38

Table II.3

Pigment (English)	Pigment (German)	Supplier	Color index		BET	Pigment/ binder	Polymeric additive							
							K	J	H	B	C	D	E	G
Brufasol yellow 1520 M	Brufasolgelb 1520 M	Bruchsaler Farbe	Y	34		4.2–5	6	19	19	26	22	17	23	25
Brufasol yellow 820 M	Brufasolgelb 820 M	Bruchsaler Farbe	Y	34		4.2–5	6	19	19	26	22	17	23	25
Brufasol yellow 880 M	Brufasolgelb 880 M	Bruchsaler Farbe	Y	34		4.2–5	6	19	19	26	22	17	23	25
Cappoxyt red 4435 B	Cappoxyt rot 4435 B	Cappelle	R	101	120		10	29	29	40	33	25	35	38
Cappoxyt red 4437 B	Cappoxyt rot 4437 B	Cappelle	R	101	95		10	29	29	40	33	25	35	38
Cappoxyt yellow 4212 X	Cappoxyt gleb 4212 X	Cappelle	Y	420	110		6	14	14	19	17	13	17	38
Carbon black FW 200	Farbruß FW 200	Degussa	Bk	7	460	0.7–0.9	9		77	105	89	67	93	175
Carbon black MA-100	Ruß MA-100	Mitsubishi	Bk	7		0.8–1	9	48	58	66	56	42	58	125
Chrome green oxide G6099	Chromoxidgruen G6099	Harcros	G	17	3		6	14	14	19	17	13	17	38
Chromofine blue 4920 GP	Chromofineblau 4920 GP	Daicolor	B	15	3		8		58	79	67	50	70	75
Chromofine blue 5080	Chromofineblau 5080	Daicolor	B	150	2 87		8		58	79	67	50	70	75

Table II.4

Pigment (English)	Pigment (German)	Supplier	Color index		BET	Pigment/binder	Polymeric additive							
							K	J	H	B	C	D	E	G
Cinquasia red Y-RT-315-D	Cinquasiarot Y-RT-315-D	Ciba-Geigy	V	19		1.3–1.8	9		58	79	66	50	70	75
Cinquasia red Y-RT-759-D	Cinquasiarot Y-RT-759-D	Ciba-Geigy	V	19		1.3–1.8	9		58	79	66	50	70	75
Cinquasia red Y-RT-985-D	Cinquasiarot Y-RT-895-D	Ciba-Geigy	V	19		1.3–1.8	9		58	79	66	50	70	75
Cinquasia violet R-RT-301 D	Cinquasiaviolett R-RT-301-D	Ciba-Geigy	V	19		0.9–1.4	9		58	79	66	80	70	75
Cinquaisa violet RT-201-D	Cinquaisaviolett RT-201-D	Ciba-Geigy	V	19		0.9–1.4	9		58	79	66	50	70	75
Copperas red R 1299	Copperas rot R 1299	Harcos	R	101		4.5–5	6	19	19	26	22	17	23	38
Copperas red R 1599	Copperas rot R 1599	Harcos	R	101		4.5–5	6	19	19	26	22	17	23	38
Copperas red R 2200	Copperas rot R 2220	Harcos	R	101			6	19	19	26	22	17	23	38
Cristal PGE 113	Cristal PGE 113	Cristal	W	6		4–5		6						
Cristal PGG 121	Cristal PGG 121	Cristal	W	6		4–5		6						

Table II.5

Pigment (English)	Pigment (German)	Supplier	Color index		BET	Pigment/ binder	Polymeric additive							
							K	J	H	B	C	D	E	G
Cristal PGM 100	Cristal PGE 100	Cristal	W	6		4–5		6						
Cristal PLG 134	Cristal PGE 134	Cristal	W	6		4–5		6						
Cromophtal red A 2 B	Cromophtalrot A 2 B	Ciba-Geigy	R	177	79	0.9–1.2	9		48	66	56	42	58	
Cromophtal red A 3 B	Cromophtalrot A 3 B	Ciba-Geigy	R	177	106	0.9–1.2	9		48	66	56	42	58	63
Elftex 125	Elftex 125	Cabot	Bk	7	30	1.3–1.8	8		38	52	44	33	47	75
Elftex 415	Elftex 415	Cabot	Bk	7	90	1.1–1.3	8		48	66	55	41	58	175
Finntitan RD 3	Finntitan RD 3	Kemira	W	6		4–5	3	6	5	7	6	4	6	5
Finntitan RDD	Finntitan RDD	Kemira	W	6		4–5	3	6	5	7	6	4	6	5
Finntitan RDE 2	Finntitan RDW 2	Kemira	W	6		4–5	3	6	5	7	6	4	6	5
Finntitan RDI-S	Finntitan RDI-S	Kemira	W	6		4–5	3	6	5	7	6	4	6	5
Finntitan RR 2	Finntitan RR 2	Kemira	W	6		4–5	3	6	5	7	6	4	6	5
Finntitan RR 2-S	Finntitan RR 2-S	Kemira	W	6		4–5	3	6	5	7	6	4	6	5
Furnace black 101	Flammruß 101	Degussa	Bk	7	20	1.5–1.8	7		38	52	44	33	47	100
Giant blue 8500	Giant blau 8500	Daihan-Swiss	B	15	3	0.9–1.3	8		58	79	67	50	70	100
Giant green 9500	Giant grün 9500	Daihan-Swiss	G	7		1.1–1.5	8		58	79	67	50	70	75

Table II.6

Pigment (English)	Pigment (German)	Supplier	Color index			BET	Pigment/binder	Polymeric additive							
								K	J	H	B	C	D	E	G
Graphtol red 2 BLS	Graphtolrot 2 BLS	Clariant	R	48	4			9		38	79	44	33	47	50
Hansa brilliant yellow 2 GX 70	Hansabrilliantgelb 2 GX 70	Hoechst	Y	74		14	0.9–1.3	7		28	38	33	25	35	75
Hansa red GG-A	Hansarot GG-A	Hoechst	O	5		12		7		28	38	33	25	35	65
Hansa yellow 10 G	Hansarot 10-G	Hoechst	Y	3		8		7		28	38	33	25	35	50
Hansa yellow G 02	Hansagelb G 02	Hoechst	Y	1				7		28	38	33	25	35	50
Hansa yellow X	Hansagelb X	Hoechst	Y	75		24		7		28	38	33	25	35	50
Heliogen blue L 6700 F	Heliogen blau L 6700 F	BASF	B	15	2	54	1.1–1.4	9		45	62	52	39	54	59
Heliogen blue L 6900	Heliogen blau L 6900	BASF	B	15	1	70	0.9–1.3	9		45	62	52	39	54	60
Heliogen blue L 6901 F	Heliogen blau L 6901 F	BASF	B	15	2	56	0.9–1.3	6		38	52	44	33	47	100
Heliogen blue L 6975 F	Heliogen blau L 6975 F	BASF	B	15	2	36	0.9–1.3	9		38	52	44	33	47	100
Heliogen blue L 7080	Heliogen blau L 7080	BASF	B	15	3	68	0.9–1.3	9		38	52	44	33	47	100
Heliogen blau L 7101 F	Heliogen blau L 7101 F	BASF	B	15	4	48	0.9–1.3	9		38	52	44	33	47	50
Heliogen green L 8605	Heliogen Grün L 8605	BASF	G	7		61	1.1–1.5	8		48	66	56	42	58	50

Table II.7

Pigment (English)	Pigment (German)	Supplier	Color index	BET	Pigment/binder	Polymeric additive							
						K	J	H	B	C	D	E	G
Heliogen green L 8690	Heliogen grün L8690	BASF	G 7	60	1.1–1.5	8		48	66	56	42	58	50
Heliogen green L 8730	Heliogen grün L8730	BASF	G 7	61	1.1–1.5	8		48	66	56	42	58	63
Heuco red 317700	Heucorot 317700	Heubach	R 177	44	0.9–1.2	9		48	66	56	42	58	63
Heucotron red 230	Heucotronrot 230	Heubach	R 104		4–5	5	15	15	21	18	13	19	20
Heucotron yellow 1070	Heucotrongelb 1070	Heubach	Y 34		4.2–5	9	14	14	19	17	13	17	
Heucotron yellow 123	Heucotrongelb 123	Heubach	Y 34		4.2–5	5	19	19	26	22	17	23	
Hombitan LW	Hombitan LW	Sachtleben	W 6	9	4–5	3	6	5	7	6	4	6	5
Hombitan R 210	Hombitan R 210	Sachtleben	W 6	17	4–5	3	6	5	7	6	4	6	5
Hombitan R 320	Hombitan R 320	Sachtleben	W 6	8	4–5	3	6	5	7	6	4	6	5
Hombitan R 505	Hombitan R 505	Sachtleben	W 6		4–5	3	6	5	7	6	4	6	5
Hombitan R 510	Hombitan R 510	Sachtleben	W 6	17	4–5	3	6	5	7	6	4	6	5
Hombitan R 522	Hombitan R 522	Sachtleben	W 6	16	4–5	3	6	5	7	6	4	6	5
Hombitan R 610D	Hombitan R 610D	Sachtleben	W 6	18	4–5	3	6	5	7	6	4	6	5
Hombitan R 611	Hombitan R 611	Sachtleben	W 6	17	4–5	3	6	5	7	6	4	6	5
Hombitan RC 566	Hombitan RC 566	Sachtleben	W 6	18	4–5	3	6	5	7	6	4	6	5

Table II.8

Pigment (English)	Pigment (German)	Supplier	Color index		BET	Pigment/binder	Polymeric additive							
							K	J	H	B	C	D	E	G
Hostaperm blue BFL	Hostapermblau BFL	Hoechst	B	15		0.9–1.3	9		48	66	56	42	58	100
Hostaperm green GG 01	Hostapermgrün GG 01	Hoechst	G	7	65	1.1–1.5	8		48	66	56	42	58	50
Hostaperm pink E 01	Hostapermrosa E 01	Hoechst	R	122	62	0.9–1.1	10		48	66	56	42	58	75
Hostaperm red E 2 B 170	Hostapermrot E 2 B 170	Hoechst	V	19	34	1.3–1.8	7		38	52	44	33	47	75
Hostaperm red E 3 B	Hostapermrot E 3 B	Hoechst	V	19	34	1.3–1.6	7		38	52	44	33	47	75
Hostaperm red E 5 B 02	Hostapermrot E 5 B 02	Hoechst	V	19	61	1.1–1.4	9		48	66	56	42	58	75
Hostaperm red P 2 GL	Hostapermrot P 2 GL	Hoechst	R	179		0.9–1.3	8		38	52	44	33	47	75
Hostaperm red P 3 GL	Hostapermrot P 3 GL	Hoechst	R	224	54	0.9–1.3	6		29	40	33	25	35	50
Hostaperm red violett ER 02	Hostapermrot-violett ER 02	Hoechst	V	19	85	0.9–1.3	8		48	66	56	42	58	75
Hostaperm scarlet GO	Hostaperm-scharlach GO	Hoechst	R	168	37	1.1–1.4	8		38	52	44	33	47	
Hostaperm violet RL special	Hostapermviolett RL spezial	Hoechst	V	23	86	0.6–0.9	10		48	66	56	42	58	75
Hostaperm violet RL-NF	Hostapermviolett RL-NF	Hoechst	V	23		0.6–0.9	10		48	66	56	42	58	

Table II.9

Pigment (English)	Pigment (German)	Supplier	Color index	BET	Pigment/binder	K	J	H	B	C	D	E	G
Hostaperm yellow H 3 G	Hostapermgelb H 3 G	Hoechst	Y 154	24	1.8–2.2			48	66	56	42	58	75
Hostaperm yellow H 4 G	Hostapermgelb H 4 G	Hoechst	Y 151	18	2–2.5	8		38	52	44	33	47	75
Hostaperm yellow H 6 G	Hostapermgelb H 6 G	Hoechst	Y 175	21		8		38	52	44	33	47	75
Irgalite blue BCA	Irgalithblau BCA	Ciba-Geigy	B 15:1	60	0.9–1.3	8		38	52	44	33	47	100
Irgalite blue GLNF	Irgalithblau GLNF	Ciba-Geigy	B 15:4		0.9–1.3	8		38	52	44	33	47	100
Irgalite blue GLVO	Irgalithblau GLVO	Ciba-Geigy	B 15:4	45	0.9–1.3	8		38	52	44	33	47	100
Irgalite red FBL	Irgalithrot FBL	Ciba-Geigy	R 48	55	1–1.4	7		28	38	33	25	35	50
Irgalite yellow F 4 G	Irgalithgelb F 4 G	Ciba-Geigy	Y 111	26	0.9–1.4	8		28	52	33	25	35	38
Irgalite yellow GO	Irgalithgelb GO	Ciba-Geigy	Y 74	17	0.9–1.3	7		28	38	33	25	35	50
Irgalite red 2 GW	Irgalithrot 2 GW	Ciba-Geigy	O 5	14	0.9–1.2	7		28	38	33	25	35	50
Irgazin red DPP-BO	Irgazinrot DPP-BO	Ciba-Geigy	R 254	16	1.3–1.8			39	53	55	42	58	100

Polymeric additive

Table II.10

Pigment (English)	Pigment (German)	Supplier	Color index	BET	Pigment/binder	Polymeric additive							
						K	J	H	B	C	D	E	G
Irgazin yellow GLTE	Irgazingelb GLTE	Ciba-Geigy	Y 109	24	0.9–1.3	8		48	66	56	42	58	63
Irgazin yellow 2 RLT	Irgazingelb 2 RLT	Ciba-Geigy	Y 110	56	0.9–1.3	7		38	52	44	33	47	
Irgazin yellow 3 RLT N	Irgazingelb 3 RLT N	Ciba-Geigy	Y 110	26	0.9–1.3	7		38	52	44	33	47	50
Krolor yellow KY-787-D	Kroloryellow KY-787-D	Cookson	Y 34		4.2–5	5	12	12	16	14	10	15	
Krolor yellow KY-795-DE	Kroloryellow KY-795-DE	Cookson	Y 34	11	4.2–5	5	12	12	16	14	10	15	
Kroma red RO 2097	Kromarot RO 2097	Harcros	R 101	15	4.5–5	6	19	19	26	22	17	23	38
Kroma red RO 3097	Kromarot RO 3097	Harcros	R 101	10	4.5–5	6	19	19	26	22	17	23	38
Kroma red RO 8097	Kromarot RO 8097	Harcros	R 101		4.5–5	6	19	19	26	22	17	23	38
Kronos 2020	Kronos 2020	Kronos	W 6		4–5	3	6	5	7	6	4	6	5
Kronos 2043	Kronos 2043	Kronos	W 6	24	4–5	3	6	5	7	6	4	6	5
Kronos 2044	Kronos 2044	Kronos	W 6	55	4–5	3	6	5	7	6	4	6	5
Kronos 2056	Kronos 2056	Kronos	W 6	11	4–5	3	6	5	7	6	4	6	5

Table II.11

Pigment (English)	Pigment (German)	Supplier	Color index	BET	Pigment/ binder	Polymeric additive							
						K	J	H	B	C	D	E	G
Kronos 2057	Kronos 2057	Kronos	W	6	17	4-5	3	6	5	7	6	4	6	5
Kronos 2059	Kronos 2059	Kronos	W	5	19	4-5	3	6	5	7	6	4	6	5
Kronos 2063	Kronos 2063	Kronos	W	6	15	4-5	3	6	5	7	6	4	6	5
Kronos 2065	Kronos 2065	Kronos	W	6	19	4-5	3	6	5	7	6	4	6	5
Kronos 2160	Kronos 2160	Kronos	W	6	15	4-5	2	6	4	6	4	3	5	13
Kronos 2190	Kronos 2190	Kronos	W	6	17	4-5	2	6	4	6	4	3	5	13
Kronos 2300	Kronos 2300	Kronos	W	6	15	4-5	3	6	5	7	6	4	6	5
Kronos 2310	Kronos 2310	Kronos	W	6	15	4-5	3	6	5	7	6	4	6	5
Kronos 2330	Kronos 2330	Kronos	W	6		4-5	3	6	5	7	6	4	6	5
Lithol fast maroon L 4763	Litholecht maroon L 4763	BASF	R	52	2	54	1.3–1.8	8		39	53	44	33	47	50
Mineral yellow 10 G 7861 X	Mineralgelb 10 G 7861 X	Cappelle	Y	34		4.2–5	5	12	12	16	14	10	15	20
Molybdate orange 6001	Molybdatorange 6001	Nubiola	R	104		4-5	6	15	15	21	18	13	19	20
Molybdate red 6055	Molybdatrot 6055	Nubiola	R	104		4-5	6	15	15	21	18	13	19	20
Molybdate red 6654	Molybdatrot 6654	Nubiola	R	104		4-5	6	15	15	21	18	13	19	20
Molybdate red 71 B	Molybdatrot 71 B	Silo	R	104		4-5	6	15	15	21	18	13	19	20

Table II.12

Pigment (English)	Pigment (German)	Supplier	Color index		BET	Pigment/binder	Polymeric additive							
							K	J	H	B	C	D	E	G
Molybdate red 72 D	Molybdatrot 72 D	Silo	R	104		4–5	6	15	15	21	18	13	19	20
Molybdate red DCC 1606	Molybdatrot DCC 1606	Dominion	R	104		4–5	6	15	15	21	18	13	19	20
Molybdate red DCC 5606	Molybdatrot DCC 5606	Dominion	R	104		4–5	6	15	15	21	18	13	19	20
Monarch 1300	Monarch 1300	Cabot	Bk	7	560	0.6–0.9			77	105	89	67	93	175
Monarch 1400	Monarch 1400	Cabot	Bk	7	560	0.6–0.9			77	105	89	67	93	175
Monarch 450	Monarch 450	Cabot	Bk	7	80	1.3–1.8	8		48	66	55	41	58	100
Monarch 800	Monarch 800	Cabot	Bk	7	220	1.1–1.3	8		48	66	55	41	58	100
Monastral green 6 Y	Monastralgrün 6 Y	Zeneca	G	36	30	1.1–1.5	8		58	79	67	50	70	75
Novoperm orange H 5 G 70	Novopermorange H 5 G 70	Hoechst	O	62	12		8		38	52	44	33	47	50
Novoperm orange HL 70	Novopermorange HL 70	Hoechst	O	36	20	1.1–1.5	9		38	52	44	33	47	75
Novoperm orange HL 70 NF	Novopermorange HL 70 NF	Hoechst	O	36		1.1–1.5	9		38	52	44	33	47	75
Novoperm red F 2 RK 70	Novopermrot F 2 RK 70	Hoechst	R	170	23	2.5–3.5	8		38	52	44	33	47	63
Novoperm red F 3 RK 70	Novopermrot F 3 RK 70	Hoechst	R	170	22	2.5–3.5	8		38	52	44	33	47	63
Novoperm red H 2 BM 01	Novopermrot H 2 BM 01	Hoechst	R	48	4	1.1–1.4	9		38	52	44	33	47	50
Novoperm red HF 3 S 70	Novopermrot HF 3 S 70	Hoechst	R	188	13	1–1.3	8		38	52	44	33	47	50

Table II.13

Pigment (English)	Pigment (German)	Supplier	Color index		BET	Pigment/ binder	Polymeric additive							
							K	J	H	B	C	D	E	G
Novoperm red HF 4 B	Novopermrot HF 4 B	Hoechst	R	187	72	1.1–1.3	8		38	52	44	33	47	50
Novoperm red violet MRS	Novopermrot-violett MRS	Hoechst	R	88	59	2–2.5	6		29	40	33	25	35	
Novoperm yellow FGL	Novopermgelb FGL	Hoechst	Y	97	26	1.1–1.4	8		38	52	44	33	47	50
Novoperm yellow HR 70	Novopermgelb HR 70	Hoechst	Y	83	22	1.2–1.6	8		38	52	44	33	47	50
Paliogen blue L 6470	Paliogen blau L 6470	BASF	B	60	52		6		29	40	33	25	35	38
Paliogen Maroon L 3920	Paliogen Maroon L 3920	BASF	R	179	52	1.1–1.3	7		48	66	56	42	58	
Paliogen red L3910 HD	Paliogen rot L3910 HD	BASF	R	178	31	0.9–1.3	8		38	52	44	33	47	50
Paliogen red violet L5080	Paliogen rotviolett L 5080	BASF	R	88	43	2–2.5	9		38	52	44	33	47	50
Paliogen yellow L1560	Paliogen gelb L 1560	BASF	Y	108	26	0.9–1.3	8		29	40	33	25	35	
Paliotol red L 3550 HD	Paliotol rot L 3550 HD	BASF	R	139		0.9–1.1	7		38	52	44	33	47	50
Paliotol yellow L 1772	Paliotol gelb L 1772	BASF	Y	153		0.9–1.3	8		38	52	44	33	47	50
Paliotol yellow L 2140 HD	Paliotol gelb L 2140 HD	BASF	Y	139	25	0.9–1.1	9		48	66	58	42	58	

Table II.14

Pigment (English)	Pigment (German)	Supplier	Color index		BET	Pigment/binder	Polymeric additive							
							K	J	H	B	C	D	E	G
Palomar blue B-4730	Palomar blue B-4730	Bayer	B 15	2	66		8		38	52	44	33	47	100
Palomar blue B-4806	Palomar blue B-4806	Bayer	B 15	1	92	0.9–1.3	8		38	52	44	33	47	100
Palomar blue B-4808	Palomar blue B-4808	Bayer	B 15	1		0.9–1.3	8		38	52	44	33	47	50
Palomar blue GB-4812	Palomar blue GB-4812	Bayer	B 15	1	80	0.9–1.3	8		38	52	44	33	47	50
Palomar green G-5420	Palomar grün G-5420	Bayer	G 36		64		8		48	66	56	42	58	50
Permanent carmine FBB 02	Permanentcar. FBB 02	Hoechst	R 146		38	0.9–1.2	8		48	66	56	42	58	63
Permanent orange RL 70	Permanentoran. RL 70	Hoechst	O 34		30		7		38	52	44	23	47	
Permanent red 2B-2311	Permananet red 2B-2311	Peer	R 48	2	74	1.1–1.3	8		38	52	44	33	47	50
Permanent red 2B-2313	Permanent red 2B-2313	Peer	R 48	20	78	1.1–1.3	8		38	52	44	33	47	
Permanent red FGR	Permanentrot FGR	Hoechst	R 112		23	1–1.2	8		38	52	44	33	47	50
Permanent red FGR 70	Permanentrot FGR 70	Hoechst	R 112		12	1–1.2	8		38	52	44	33	47	50

Table II.15

Pigment (English)	Pigment (German)	Supplier	Color index	BET	Pigment/binder	Polymeric additive							
						K	J	H	B	C	D	E	G
Perrindo red R-6418	Perrindo red R-6418	Bayer	R 224		0.9–1.3	7		29	40	33	25	35	38
Phthalo blue 3300	Phthalo blue 3300	Peer	B 15	58		8		48	66	56	42	58	100
Phthalo blue 4201	Phthalo blue 4201	Peer	B 15:2	58		8		48	66	56	42	58	100
Phthalo blue 8530	Phthalo blue 8530	Peer	B 15:4	55		8		48	66	56	42	58	100
Phthalo green 9501	Phthalo grün 9501	Peer	G 7	62		9		58	79	67	50	70	75
Printex 45	Printex 45	Degussa	Bk 7	90	1.4–1.8	8		48	66	56	42	58	
Printex G	Printex G	Degussa	Bk 7	30	1.1–1.5	8		29	40	33	25	35	50
Quindo magenta RV-6830	Quindo magenta RV-6830	Bayer	R 202	98				38	52	44	33	47	
Quindo magenta RV-6832	Quindo magenta RV-6832	Bayer	R 122		0.9–1.1	9		58	79	67	50	70	70
Quindo red RV-6700	Quindo red RV-6700	Bayer	V 19			8		38	52	44	33	47	50
Quindo violet RV-6902	Quindo violet RV-6902	Bayer	V 19	59		8		58	79	67	50	70	75
Quindo violet RV-6926	Quindo violet RV-6926	Bayer	V 19	78		8		58	79	67	50	70	75
Raven 1035	Raven 1035	Columbian	Bk 7	95		7		38	52	44	33	47	

Table II.16

Pigment (English)	Pigment (German)	Supplier	Color index	BET	Pigment/ binder	Polymeric additive							
						K	J	H	B	C	D	E	G
Raven 14	Raven 14	Columbian	Bk 7	24		7		38	52	44	33	47	
Raven 2000	Raven 2000	Columbian	Bk 7	120		8		48	66	56	42	58	75
Raven 420	Raven 420	Columbian	Bk 7	25		6		38	52	44	33	47	
Raven 5000	Raven 5000	Columbian	Bk 7	430		9		58	79	67	50	70	175
Sandorin blue BNLF	Sandorinblau BNLF	Clariant	B 15	2	0.9–1.3	9		38	52	44	33	47	100
Sandorin green 8 GLS	Sandoringrün 8 GLS	Clariant	G 36	37	1.1–1.5	8		48	66	56	42	58	50
Sandorin red violet 3 RL	Sandorinrotvio. 3 RL	Clariant	R 257		0.7–1	7		48	66	56	42	58	
Sandorin violet BLP	Sandorinvio. BLP	Clariant	V 23	85	0.6–0.9	8		48	66	56	42	58	63
Sandorin yellow 5 GD	Sandoringelb 5 GD	Clariant	Y 173		0.9–1.2	8		34	47	39	29	41	44
Sico fast orange L 3052 HD	Sicoecht orange L 3052 HD	BASF	O 5	10	1.1–1.4	7		29	40	33	25	35	50
Sico fast scarlet L 4252	Sicoecht scharlach L 4252	BASF	R 170	23	2.5–3.5	8		38	52	44	33	47	63
Sico fast yellow NBD 1752	Sicoecht gelb NBD 1752	BASF	Y 83			8		38	52	44	33	47	50
Sico yellow L 0951	Sico gelb L 0951	BASF	Y 42			10	29	29	40	33	25	35	38
Sico yellow L 1252 HD	Sico gelb L 1252 HD	BASF	Y 74	16	0.9–1.3	7		29	40	33	25	35	38
Sicomin red L 3030 S	Sicomin rot L 3030 S	BASF	R 104	11	4–5	5	12	12	16	14	10	15	

Table II.17

Pigment (English)	Pigment (German)	Supplier	Color index		BET	Pigment/ binder	Polymeric additive							
							K	J	H	B	C	D	E	G
Sicomin red L 3125	Sicomin rot L 3125	BASF	R	104	12	4–5	6	10	10	14	11	8	12	
Sicomin yellow D 1122	Sicomin gelb D 1122	BASF	Y	34	10	4.2–5	5	12	12	16	14	10	15	
Sicomin yellow FC 1225	Sicomin gelb FC 1225	BASF	Y	34	4	4.2–5	5	12	12	16	14	10	15	
Sicomin yellow L 1522	Sicomin gelb L 1522	BASF	Y	34	22	4.2–5	5	12	12	16	14	10	15	
Sicomin yellow L 1635 S	Sicomin gelb L 1635 S	BASF	Y	34	8	4.2–5	5	12	12	16	14	10	15	
Sicotan yellow L 1011	Sicotan gelb L 1011	BASF	Y	34		4.2–5	5	12	12	16	14	10	15	
Sicotan yellow L 1012	Sicotan gelb L 1012	BASF	Y	53	3	4–5	5	19	19	26	22	17	23	25
Sicotan yellow L 1910	Sicotan gelb L 1910	BASF	Br	24	5	4–5	6	19	19	26	22	17	23	38
Sicotrans red L 2816	Sicotrans rot L 2816	BASF	R	101	100		10	38	38	52	44	33	47	63
Sicotrans red L 2817	Sicrotrans rot L 2817	BASF	R	101	93		10	29	29	40	33	25	33	38
Sicrotrans yellow L 1915	Sicotrans gelb L 1915	BASF	Y	42	90		10	29	29	40	33	25	35	38
Sicotrans yellow L 1916	Sicrotrans gelb L 1916	BASF	Y	42	80		10	29	29	40	33	25	35	38
Special black 100	Spezial schwarz 7	Degussa	Bk	7	30	1.3–1.8	10		58	79	67	50	70	

Table II.18

Pigment (English)	Pigment (German)	Supplier	Color index			BET	Pigment/ binder	Polymeric additive							
								K	J	H	B	C	D	E	G
Special black 4	Spezial schwarz 4	Degussa	Bk	7		180	1.2–1.6	8		48	66	56	42	58	
Special black 6	Spezial schwarz 6	Degussa	Bk	6		300	1.1–1.4			77	105	89	67	93	100
Sunbrite red 234-0071	Sunbriterot 234-0071	Sun	R	480	20	55		7		48	66	56	42	58	
Sunfast blue 249-0592	Sunfastblau 249-0592	Sun	B	15	4	77		8		45	66	52	39	54	60
Sunfast blue 249-1282	Sunfastblau 249-1282	Sun	B	15	3	72		8		45	66	52	39	54	60
Sunfast green 7 264-0414	Sunfastgrün 7 264-0414	Sun	G	7		66		8		48	66	55	41	58	
Sunfast green 7 264-7117	Sunfastgrün 7 264-7117	Sun	G	7		47		8		48	66	55	41	58	
Sunfast yellow 14 274-0686	Sunfastgelb 14 274-0686	Sun	Y	140		36		7		38	52	44	33	47	
Sunfast yellow 14 274-3954	Sunfastgelb 14 274-3954	Sun	Y	14		31		7		38	52	44	33	47	
Symular red 3037	Symularrot 3037	DIC	R	48	4		1.1–1.5	8		48	66	56	42	58	
Thiosol GBL	Thiosol GBL	Cappelle	R	104			4–5	5	19	19	26	22	17	23	25
Ti-Pure R 706	Ti-Pure R 706	Du Pont	W	6			4–5	3	6	5	7	6	4	6	5
Ti-Pure R 900	Ti-Pure R 900	Du Pont	W	6			4–5	3	6	5	7	6	4	6	5
Ti-Pure R 901	Ti-Pure R 901	Du Pont	W	6			4–5	3	6	5	7	6	4	6	5

Table II.19

Pigment (English)	Pigment (German)	Supplier	Color index	BET	Pigment/ binder	Polymeric additive								
						K	J	H	B	C	D	E	G
Ti-Pure R 902	Ti-Pure R 902	Du Pont	W	6	15	4-5	3	6	5	7	6	4	6	5
Ti-Pure R 931	Ti-Pure R 931	Du Pont	W	6		4-5	3	6	5	7	6	4	6	5
Ti-Pure R 960	Ti-Pure R 960	Du Pont	W	6		4-5	3	6	5	7	6	4	6	5
VHG	VHG													
Tiona RCL-376	Tiona RCL-376	SCM	W	6	32	4-5	3	6	5	7	6	4	6	5
Tiona RCL-388	Tiona RCL-388	SCM	W	6	25	4-5	3	6	5	7	6	4	6	5
Tiona RCL-472	Tiona RCL-472	SCM	W	6	12	4-5	3	6	5	7	6	4	6	5
Tiona RCL-535	Tiona RCL-535	SCM	W	6	11	4-5	3	6	5	7	6	4	6	5
Tiona RCL-6	Tiona RCL-6	SCM	W	6		4-5	3	6	5	7	6	4	6	5
Tiona RCL-628	Tiona RCL-628	SCM	W	6	18	4-5	3	6	5	7	6	4	6	5
Tiona RCL-666	Tiona RCL-666	SCM	W	6		4-5	3	6	5	7	6	4	6	5
Tiona RCL-9	Tiona RCL-9	SCM	W	6		4-5	3	6	5	7	6	4	6	5
Tioxide R-HD-2	Tioxide R-HD-2	Tioxide	W	6		4-5	3	6	5	7	6	4	6	5
Tioxide R TC-2	Tioxide R-TC-2	Tioxide	W	6		4-5	3	6	5	7	6	4	6	5
Tioxide R-TC-30	Tioxide R-TX-30	Tioxide	W	6		4-5	3	6	5	7	6	4	6	5
Tioxide R-TC-4	Tioxide R-TC-4	Tioxide	W	6		4-5	3	6	5	7	6	4	6	5
Tioxide R-TC-90	Tioxide R-TC-90	Tioxide	W	6		4-5	3	6	5	7	6	4	6	5
Tioxide R-XL	Tioxide R-XL	Tioxide	W	6		4-5	3	6	5	7	6	4	6	5
Tioxide TR 50	Tioxide TR 50	Tioxide	W	6	25	4-5	3	6	5	7	6	4	6	5
Tioxide TR 61	Tioxide TR 61	Tioxide	W	6		4-5	3	6	5	7	6	4	6	5

Table II.20

Pigment (English)	Pigment (German)	Supplier	Color index	BET	Pigment/binder	Polymeric additive							
						K	J	H	B	C	D	E	G
Tioxide TR 63	Tioxide TR 63	Tioxide	W 6		4–5	3	6	5	7	6	4	6	5
Tioxide TR 80	Tioxide TR 80	Tioxide	W 6		4–5	3	6	5	7	6	4	6	5
Tioxide TR 92	Tioxide TR 92	Tioxide	W 6	14	4–5	2	10	10	14	4	3	5	5
Tipaque CR-50	Tipaque CR-50	Ishihara	W 6		4–5	3	6	5	7	6	4	6	5
Tipaque CR-50-2	Tipaque CR-50-2	Ishihara	W 6		4–5	3	6	5	7	6	4	6	5
Tipaque CR-58	Tipaque CR-58	Ishihara	W 6		4–5	3	6	5	7	6	4	6	5
Tipaque CR-58-2	Tipaque Cr-58-2	Ishihara	W 6		4–5	3	6	5	7	6	4	6	5
Tipaque CR-80	Tipaque CR-80	Ishihara	W 6		4–5	3	6	5	7	6	4	6	5
Tipaque CR-90	Tipaque CR-90	Ishihara	W 6		4–5	3	6	5	7	6	4	6	5
Tipaque CR-93	Tipaque CR-93	Ishihara	W 6		4–5	3	6	5	7	6	4	6	5
Tipaque CR-97	Tipaque CR-97	Ishihara	W 6		4–5	3	6	5	7	6	4	6	5
Tipaque R-930	Tipaque R-930	Ishihara	W 6		4–5	3	6	5	7	6	4	6	5
Toyo RF 49802	Toyo RF 49802	Toyo Ink	B 15	2	1.1–1.3	7		38	52	44	33	47	100
Tronox CR 800	Tronox CR 800	Kerr-McGee	W 6		4–5	3	6	5	7	6	4	6	5
Tronox CR 800 PG	Tronox CR 800 PG	Kerr-McGee	W 6		4–5	3	6	5	7	6	4	6	5
Tronox CR 813	Tronox CR 813	Kerr-McGee	W 6		4–5	3	6	5	7	6	4	6	5
Tronox CR 821	Tronox CR 821	Kerr-McGee	W 6		4–5	3	6	5	7	6	4	6	5
Y-933-D	Y-933-D	Cookson	Y 34	8		5	12	12	16	14	10	15	

Table II.21

Pigment (English)	Pigment (German)	Supplier	Color index		BET	Pigment/ binder	Polymeric additive							
							K	J	H	B	C	D	E	G
Y-969-D	Y-969-D	Cookson	Y	34	9		5	12	12	16	14	10	15	
YE-937-LD	YE-937-LD	Cookson	R	1040	12		6	10	10	14	11	8	12	
YE-998-LD	YE-998-LD	Cookson	R	1040	16		6	10	10	14	11	8	12	

Appendix III: Special Notes on Surface Measurement Techniques: Transparency, Gloss, Haze, and Color

What Influences Transparency?

The appearance of a transparent product is determined by dozens of factors—such as viewing conditions, material selection and process parameters. (Please see Figure III.1 for a depiction of basic viewing dynamics.) Homogenous materials with smooth surfaces may, under the proper conditions, appear glossy and crystal clear. In contrast, diffused scattering caused by pigments or rough surfaces can significantly reduce appearance quality. Accordingly, a comprehensive list of essential evaluation criteria must include transmittance, haze, and clarity.

Factors Influencing Transparency

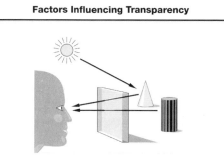

Figure III.1 Viewing conditions, material selection, and process parameters

How Are Objects Visually Evaluated?

Since a picture is indeed worth a thousand words, especially when one is discussing visual appearance, this appendix will employ photographs and/or diagrams wherever possible—beginning with the following dice example.

If objects are alternately viewed through two transparent specimens demonstrating different light-scattering characteristics, then the two resultant images may be completely different. For instance, "wide angle scattering" equally diffuses the scattered light in all directions, resulting in a milky/hazy appearance and a

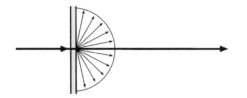

Figure III.2 Wide angle scattering

Figure III.3 Narrow angle scattering

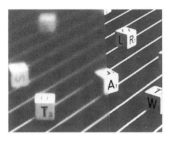

Figure III.4 Comparative scattering (wide)

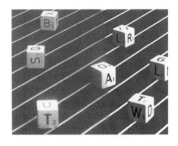

Figure III.5 Comparative scattering (narrow)

demonstrable loss-of-contrast (Fig. III.2). In comparison, narrow angle scattering (Fig. III.3) deflects the light in small angles, therefore concentrating the scattered light in a narrow cone. Overall haziness is markedly different from the previous example. (Actual photographs of both wide angle and narrow angle scattering are shown in Figures III.4 and III.5.)

The criteria used to distinguish narrow angle scattering from wide angle scattering are, to some extent, distance dependent. As the distance between an object and a transparent material increases, the "see-through quality" (sometimes denoted as "contact clarity") decreases, whereas the haze remains relatively constant.

How are Haze and Transparency Measured?

When measuring haze, a controlled light beam strikes the specimen and enters an integrating sphere (an example measuring apparatus is shown in Figure III.6). The transmitted light is equally diffused by the special matt-finish white coating of the sphere's interior surface. Total transmittance is determined with the sphere exit port closed, while haze is measured with the sphere exit port open. A ring sensor precisely mounted at the exit port of the sphere detects narrow angle scattered light (clarity).

Haze Detection Apparatus

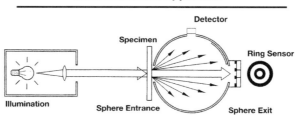

Figure III.6 Schematic diagram

In summary, the objective measurement of total transmittance, haze, and clarity helps guarantee uniform and consistent product quality. One should always closely monitor the influence of both process parameters and material properties on optical properties. Total appearance is often influenced not only by the flow chart variables mentioned in Figure III.7, but also by the following:

Process Parameters
- Temperature (especially germane with extrudates)
- Mass homogeneity
- Temperature control in the processing equipment
- Cooling rate
- Die geometry/roller surface
- Flow defects
- Film thickness

Appearance Flowchart

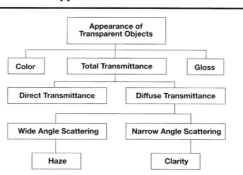

Figure III.7 Flowchart

Material Properties
- Molecular structure
- Molecular mass distribution
- Additives
- Compatibility
- Rheology

Pigment Color and Perception

Color perception, an important aspect of our five senses, is strongly influenced by external factors such as object size, ambient color, and lightness, as well as personal factors—including gender, age, fatigue, and even the mood of the observer. Color researchers, as well as behavioral scientists, have often noted that human color perception is often plagued by the following difficulties:

Subjective Observations
- Little or no recollection of colors or color differences
- Difficult and vague communication of color phenomena

Such difficulties are best eliminated by using color instrumentation and internationally specified color systems which guarantee objective description of color specimens and their concomitant color differences. As a beginning step in the direction of objectivity, color perception is often defined as being dependent on the measured interrelationships existing among the three factors shown in Figure III.8. The spectral functions of light source and observer are delineated, for instance, by the CIE (Commission International d'Éclairage) and can be stored in the memory of a color instrument that allows reproducible color measurement. Within the instrument, all three factors are specially integrated to define specific and reproducible colors.

With smooth, high gloss surfaces, the light reflected generally follows classical reflection laws; accordingly, the intensity of the reflected light is calculated from the refractive index of the material using the Fresnel equation (Fresnel-Reflection).

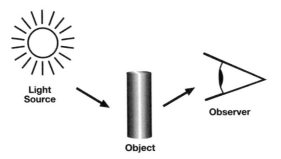

Figure III.8 Three factors influencing color perception

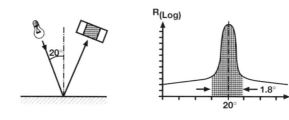

Figure III.9 Gloss measurement

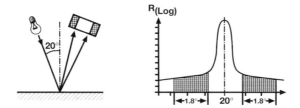

Figure III.10 Haze measurement

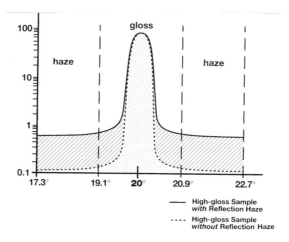

Figure III.11 High-gloss samples

When measuring gloss (reflectometer value), the intensity of light reflected from a surface is measured in an angle range demarcated by the aperture dimensions of the instrument (and also related to the light reflected from a primary luminous standard). For instance, the angle range of a 20° geometry is ±0.9, i.e. 1.8° (Fig. III.9).

For the measurement of reflection haze, two additional detectors are positioned on either side of the aperture. This enables the instrument to detect the scattered light intensity (Fig. III.10), which is also related to a standard. In general, the measurement results are displayed logarithmically. Figure III.11 displays composite haze and gloss measurements; additional background information (regarding color measurement and control) is shown also in Figure III.12.

Important Variables for Color Measurement and Control

1. Calibration
- Prerequisite for accurate results
- Traceable white and black standards
- Clean, dust- and scratch-free standards

2. Sample preparation
- Fingerprint-, dust- and scratch-free surfaces
- Instrument port completely covered
- Consistent and documented procedure

3. Curved samples
- Instrument port completely covered
- Always measure the same radius

4. Structured samples
- Take measurements at several areas and calculate the mean value
- For highly structured samples – average readings taken in standard directions A, B, C, and D
- Consistent and documented procedure

5. Transparent, low chroma samples
- Uniform film thickness
- Consistent, stable backing material (e.g. ceramic tiles)
- Fold films several times or measure multiple layers
- Consistent and documented procedure

6. Yellowing
- Use Yellowness/Whiteness Indices to determine weathering, aging, light and temperature influences

7. Hiding power
- Use opacity to determine yield and hiding power of paints, transparent films and plastics.

Procedure:
- Apply uniform film thickness on black/white contrast chart
- Measure over black and white background

$$\text{Opacity (\%)} = \frac{\text{Y Black}}{\text{Y White}} \times 100 \ (\%)$$

100% = complete hiding

8. Light-source change
- Metamerism: two specimens match under one illuminant, but do not match under another one
- This color difference can be caused by using different types of colorants
- Metameric pairs have spectral curves that cross at least three times
- Practical setting: metamerism index MI

MI $\geq$ 1 mismatch
MI < 1 match

Figure III.12 Additional background information (regarding color measurement and control)

Appendix IV: Unique Challenges in UV/EB-Curing Systems

Overview

The evolutionary advance of UV/EB-curing systems has created a new realm of technology in which the demands for performance improvement are more stringent than ever. Furthermore, *performance enhancement, in and of itself, is simply not enough; the simultaneous achievement of quality improvement* has become essential to winning the typical customer's loyalty. An in-depth discussion of newly patented additives will be presented within the dual theoretical and practical framework of the following crucial issues:

- Levelling, flow, and defoaming/air release
- The control of flatting behavior

Levelling, Flow, and Defoaming/Air Release: Theory and Application

Generally speaking, the vast majority of flow and levelling additives can be classified as either "siloxanes" or "acrylates". Regardless of which additive category is chosen, one must carefully consider both the viscosity and the solids level of the coating system at hand. Obviously, the aforementioned decision criteria are applicable to virtually all coatings systems, including those which are not UV/EB-curing in nature.

As a rule of thumb—high viscosity, high-solids systems require levelling additives; in contrast, low viscosity, low-solids systems require "surface flow control" (SFC) additives. (Please note that, in regard to semantics and nomenclature, several variant "SFC terms and expressions" are utilized in different industry circles. From a practical standpoint, though, the technical jargon should not serve as a stumbling block to understanding the phenomena at hand.) For instance, what exactly are the distinguishing characteristics of "levelling" versus "SFC" products, especially in UV/EB systems? What are the physical and chemical determinants of flow and levelling improvement? The provision of answers to the above questions will serve as the focal point of Appendix IV.

Levelling

Suppose that one observes the surface behavior of a typical high viscosity system at two discrete points in time—*first, immediately after application, and then immediately after curing.* What can one learn about levelling behavior? Quite unsurprisingly, the observed system will generally display a highly structured, somewhat uneven surface directly after application. Then the curing process begins, and the next obvious question becomes whether or not this rather imperfect surface will be able to sufficiently level itself to the point where all surface defects disappear. From an interfacial tension perspective, one must transform a relatively immobile air/liquid interface into an exceedingly mobile air/liquid interface. In coatings systems which have not been fully optimized, the dynamics of the system simply prevent proper levelling; the end result is, of course, a rather imperfect surface even after the curing process has been consummated.

From the perspective of the formulator, what can be done to prevent levelling problems? Special levelling additives (polyacrylates and/or acrylic copolymers) can be utilized to completely smooth and equilibrate surface features after application. Interestingly enough, certain acrylic levelling additives can, in selected instances, slightly *elevate* the coating system's surface tension. *Since one of the basic laws of liquid phase physics states that all liquid entities seek to stabilize themselves by occupying as little volume (per given surface area) as possible, then obviously all "micro-peaks" and "micro-valleys" will disappear.* Proper levelling is the result.

Surface Flow Control

As mentioned previously, "levelling" phenomena occupy center stage in regard to *high* viscosity systems; however, "surface flow control" phenomena are most crucial in *low* viscosity systems. Once again, observations at two discrete points in time are essential.

Immediately after application, a typical low viscosity system remains mobile enough to form a smooth, even surface. What happens next though? Because of the curing process itself, localized surface tension differentials and eddy currents begin to interrupt the surface integrity. The aforementioned disturbances persist throughout the curing process, with the end result being the same well-known surface imperfections discussed above.

In this case, the most optimal formulating option is the utilization of siloxane-based additives to actually *lower* the surface tension of the entire coating system. Siloxanes therefore provide the key to proper surface flow control during the crucial curing phase of low viscosity systems. *Performance improvement with SFC additives, in stark contrast to improvement with levelling additives, is absolutely never dependent upon the production of surface tension increases.* Additional reasons for the employment of SFC additives are shown below:

- Improved flatting agent orientation
- Homogeneous flatting agent distribution
- Optimized substrate wetting
- Elimination of Bénard cells
- Avoidance of air draft sensitivity
- Improved surface slip properties

Newly patented SFC functionalities include UV-reactive moieties ($-CH=CH_2$), special aralkyl groups and modified polyether side-chains. One reason that polysiloxanes are so effective is that they reduce the surface tension of the paint system, thereby equalizing local surface tension differences during the drying period.

At the same time, this reduction in surface tension also enables the coating system an opportunity to wet the substrate. Traditional dimethylpolysiloxanes and methylalkylpolysiloxanes can be employed in certain aqueous systems (where the non-volatile component is at least 15%) and in nearly all solvent-based systems. Shorter alkyl groups exert greater influence upon surface tension reduction.

Not only do dimethylpolysiloxanes impart more surface tension reduction than methyl-alkylpolysiloxanes, but they also impart more sensitivity to "touch-up," repair, and/or surface cleaning operations.

In aqueous UV dispersions, for instance, the binders are present in a rather fixed macromolecular state, thus the defoaming mechanism that is most often successful in solvent-based systems—*controlled incompatibility of the defoamer with the binder*—cannot adequately defoam most aqueous systems. As a result, the goal of removing foam from environmentally friendly wood and furniture coatings is often highly challenging. (This discussion is not, however, limited to systems for the wood and furniture coatings industry.) Depending on the type and quantity of the emulsifying agents used in the binder system, the defoamer must be either more hydrophilic or more hydrophobic. For instance, paint systems containing acrylics as binders, due to the lamella-enhancing emulsifying agents used, tend to display severe foam stabilization, so especially effective defoamers are often necessary.

Polyurethane binder systems exhibit less tendency to foam because they are produced free of emulsifying agents. Furthermore, if and when foam occurs in such systems, it can be destroyed more quickly and easily. On the other hand, polyurethane systems may be more prone to crater formation when the wrong defoamer is selected.

The active ingredients of defoamers and air release additives typically exhibit a positive spreading coefficient, and such ingredients are often based on silicones, polyglycols, and/or polyureas. Selection of the most suitable defoamer or air release additive depends, in part, upon the binder system. For example, acrylic dispersions require hydrophobic defoamers; PU-dispersions require hydrophilic defoamers. More importantly though, the selection depends upon the shear forces encountered and upon the viscosity of the paint system. Practical experience shows that (depending upon the shear forces available), more than one additive may be required to sufficiently defoam the system.

A comparative example (with a typical acrylic system versus a polyurethane) will now be described to illustrate the crucial impact of the resin system choice. First of all, in regard to the binders employed in this particular example, the acrylic contains a high level of emulsifying agents, and therefore proves to be more difficult to defoam than the PU-dispersion. Next, in regard to the shear forces available, the pigment grinding phase for the acrylic formulation involves higher shear forces and therefore greater foam stabilization tendency.

How do the aforementioned resin-specific differences impact defoamer usage?

In the millbase of the pigmented system, a strongly hydrophobic defoamer can be used once sufficient shear forces are present. During later steps in the paint production and application procedure, the shear forces are reduced and therefore become insufficient for incorporating the defoamer. As a result, a relatively hydrophilic defoamer must be selected.

In an unpigmented low viscosity PU-dispersion system (without emulsifiers), a hydrophilic defoamer should be utilized from the very beginning due to the absence of high shear forces. This obviously helps avoid the possibility of cratering that might arise from the usage of strongly hydrophobic defoamers. If, however, a hydrophilic defoamer is *incorporated under high shear in the millbase*, then the defoamer will be emulsified and will, as a result, lose its efficiency.

In contrast to conventional polysiloxanes, special short-chain backbone "silicone surfactants" can be employed in aqueous systems containing less than 15% co-solvent. One very important differentiating characteristic of silicone surfactants is the absence of surface slip increase. Of course, combinations of both polysiloxanes and silicone surfactants can be employed where the dual performance features of both may be required.

Solvent Free UV- and EB-Curing Systems: Formulating Hints

For the formulation of solvent-free UV- and EB-curing systems, one can employ acrylic oligomers (urethane-, epoxy-, and polyester-acrylates) as binders. To adjust viscosity, acrylic monomers (for example, di-functional and tri-functional acrylics) are preferable. It is not necessarily essential to always know whether the system is EB-versus UV-curing, since the respective systems often differ from one another only in the quantity of photoinitiator employed.

Differentiation between defoaming and air release is absolutely necessary for UV- and EB-curing systems. Foam may obviously occur not only within the paint film itself—but also at the surface. Viscosity is of prime importance. If the system exhibits high viscosity, then air release is most significant because foam bubbles move slowly to the surface. In low viscosity systems, however, foam bubbles can reach the paint surface relatively quickly.

In solvent-free systems, poor levelling can also present problems—especially upon roller application. Acrylic additives are the best choices for levelling improvement in such formulations. To improve surface slip, a suitably compatible and modified polysiloxane (which reduces surface tension) should be incorporated.

The Control of Flatting Behavior

During the past few decades, synthetic silicas have performed an integral role in the production of "flat" or non-glossy coatings. Recent chemical advances allow significantly increased control of flatting behavior through the utilization of special acidic-functionality copolymeric additives.

Flatting Behavior Control in "Traditional" Systems

Before an in-depth discussion of performance improvement mechanisms can be presented, a brief mention of flatting behavior in traditional (non-UV/EB) systems is necessary. Perhaps the best and most succinct approach is the provision of an answer to the following question: *Exactly which physical phenomena are responsible for supporting the "flatting action" of the most commonly employed flatting agents— synthetic silica products?* In general, the evaporation of solvent or water leads to shrinkage of the coating film. This allows the flatting agents the possibility of reaching the surface where they, in turn, can perform the vital function of lowering the gloss. In addition to the aforementioned "shrinkage effect", an important second factor—the presence of high molecular-weight polymeric resin moieties—can dramatically enhance flatting behavior.

Flatting Behavior Control in UV/EB-Systems

In comparison to "traditional" coatings, UV/EB-systems often present dual challenges. Flatting behavior control can *first* be severely hampered by the absence of solvents and *then* can be even further hampered by the absence of high molecular-weight resin moieties. (Please note that the definition of "solvents" can be expanded, as appropriate, to include water as a solvating medium.) Proper orientation of the flatting agent may furthermore be adversely affected to the point where both reproducibility and efficiency of flatting behavior are far from optimal. It is not uncommon for the resultant problems to generate batch-rejection rates of 30 to 40% and raw materials cost overruns of 20 to 30%. Dramatic performance improvements —*including reproducible lowering of both gloss (by up to 20 units) and viscosity (by up to 25% or more)*—have been achieved through utilizing copolymeric additives.

Background Information: A Comparative Overview of Sub-Optimal Flatting Control Methods for UV/EB-Systems

Of course, some UV/EB-systems exhibit improved flatting control when "excessively" high flatting agent levels are employed; nevertheless, an important trade-off (viscosity increase) nearly always occurs.

Another popular performance enhancement method involves the devotion of exceedingly close attention to the evaluation and usage of a tailored flatting agent. Everyone recognizes the fact that not all fumed silicas and related products are created equal; however, exactly *how much influence can be exerted* by tailored flatting agents? Two "rules-of thumb" are mentioned below:

1. UV/EB-systems which contain *polyester/styrene* functionalities are more amenable to usage with special flatting agents displaying *larger particle sizes* and *no organic surface treatments*.
2. UV/EB-systems which contain *acrylate/acrylic* functionalities are preferably used with special flatting agents displaying *smaller particle sizes* and *copious organic surface treatments*.

In spite of the apparent utility of the above "stop-gap" methods, several rather serious side-effects and repercussions may arise. Increased viscosity, reduced "workability", sporadic foam/air entrapment, and sub-optimal flow and levelling may occur.

The Key to Optimal Flatting Control

As mentioned previously—acidic functionality copolymeric additives can provide the key to performance improvement. Such additives *completely wet virtually all micro-surfaces of the flatting agent molecules.*

Micrographs of UV/EB-systems containing copolymeric additives demonstrate the *absence* of agglomerate structures and the *presence* of either deflocculated primary particles or specially controlled (and deflocculated) "aggregate structures". *The viscosity results are even more unequivocal; viscosity reductions of up to 25% have been achieved with additive usage levels of only 2 to 35%* (solids based upon the weight of the flatting agent). Raw material utilization, particularly in regard to the flatting agent itself, is optimized to the point where less than half the previously required quantity may be necessary (since unwetted synthetic silica surfaces no longer exist). In addition, one can still exercise the option of utilizing more flatting agent (in conjunction with the additive) to *enhance* certain properties or perhaps even to *increase* the "observed degree of flatness" beyond previously achievable levels.

With the express objective of demonstrating, on a practical basis, the nature of additive-induced flatting behavior control—several experiments were performed. Syloid ED30 and Syloid ED50 (67% larger particle size than ED30) were employed in the production of flatting agent pastes. The level of acidic-functionality copolymeric additive varied from 2% to 35%. The experimental regimen included a wide variety of tests; however, special emphasis was placed upon viscosity measurements. Given the logical requirements of the test battery, along with the idiosyncrasies of fluid systems containing flatting agents, a cone-and-plate device was chosen for many tests. (The above device often provided the most complete characterization of viscosity and viscosity-related phenomena.) A synopsis of the more pertinent experimental results is shown in Table IV.1.

Table IV.1   Additive-Induced Flatting Behavior Control: Sypnosis of Experimental Results

- In flatting pastes with Syloid ED30, it became evident that appropriate *viscosity* measurements *could not even be performed under the prevailing test conditions unless at least 5% copolymeric additive A* (solids based on flatting agent quantity) *were present.* Upon adjusting the additive level from 5 to 15%, *progressive viscosity decreases of as much as 50% were noted*

- Comparatively speaking, the laboratory results with Syloid ED50 pastes demonstrated significantly steeper viscosity reduction curves. In fact, the initial viscosity reduction began at silica usage levels below 0.25%; peak performance was reached at 5%

- From both logical and empirical viewpoints, the product-specific correlation of performance enhancement to silica usage levels is, of course, dependent upon the disparate flatting agent particle sizes. (More polymeric additive is required to deflocculate the *larger* composite surface area of *smaller* particles.)

- Additional experiments with coating systems containing various levels of flatting agents were performed; selected results will now be discussed. Several test series focused on Syloid ED30 since this product's small particle size and *organic treatment* predispose it to usage in UV/EB systems. An optimal additive level of 15% was employed in a comparative test battery in which flow and viscosity curves were measured   *In terms of final results, viscosity reductions (versus the "additive-free" system) of greater than 25% were noted even in systems containing enough flatting agent to demonstrate gloss levels of less than 40.* From a practical standpoint, this means that flatting behaviour can easily be optimized; flatting agent levels (and the concomitant increases in degree of flatness) can be raised while maintaining constant viscosity

- When one observes the individual flow thresholds of the systems described, then one discovers the additional advantage of lower flow thresholds as a direct corollary of copolymeric additive usage. The resultant improvement in surface properties is both economically and aesthetically desirable

- The aforementioned experiments have proven the utility of copolymeric additive A in effectively decreasing gloss by as much as 20 units (even in systems which already have absolute gloss levels as low as 40)

Conclusions

Performance enhancement in UV/EB-curing systems presents multiple challenges in regard to flow and levelling; defoaming and air release; and flatting behavior. Optimal solutions require not only performance enhancement—but also *cost-effective quality improvement*. Accordingly, three classes of additives provide the "missing link" which can satisfy all crucial constraints encountered by the formulator of even the most advanced UV/EB-curing systems.

- SFC-additives and modified siloxanes (for the tailored control of flow and/or levelling characteristics)

- Lamella-disrupting polymers and siloxane derivatives (for the destruction of foam and/or entrapped air)
- Special acidic-functionality copolymers (for the provision of flatting behavior control)

Special Note: Because of standard industry conventions, some alphabetic and/or numeric codes (including, for instance, the "copolymeric additive A" designation) were required to be used, in some cases, for multiple references. Accordingly, in situations where the reader may be interested in augmented information, please contact the publisher or author.

Appendix V: The Comparative Interdependence of Thermostability and Intercoat Adhesion

Synopsis

Modern polysiloxane chemistry serves an integral role in the reduction of interfacial tension; nevertheless, not all siloxanes are created equal. What accounts for the differential temperature stability traits exhibited by various polysiloxane structures? What are the determinants of proper intercoat adhesion? More importantly, how can one simultaneously *enhance* performance, *reduce* interfacial tension, and *avoid* possible deleterious side-effects? Furthermore, how can one achieve the above on a consistent and reproducible basis?

In response to the above questions, the following crucial issues are overviewed:

- Review of introductory siloxane chemistry
- The comparative thermostability of polyether-modified versus polyester-modified polysiloxanes
- Intercoat adhesion effects
- Conclusions and practical implications

Introductory Siloxane Chemistry

Silicones are important performance-enhancing ingredients; in fact, the stringent quality demands placed upon modern coatings often necessitate the utilization of silicones. Although literally dozens of performance enhancements can be made possible, the following three enhancements often serve as the major rationale for silicone usage:

- Reduction of surface tension within the overall coating system (In this fashion, improved wetting of "challenging" substrates becomes possible—even when highly polar resins are employed.)
- Improvement of scratch resistance as a result of reduced friction values
- Improvement of flow/levelling properties (especially in regard to coatings systems applied by some of the more modern techniques)

The most commonly employed silicones in the coatings industry are, semantically speaking, classified as polyether-modified dimethylpolysiloxanes. As a matter of introduction and review, the two most important structural elements of typical products are delineated below:

- *Dimethyl units* which accentuate surface tension reduction.
- *Polyether chains* which enhance paint system compatibility and/or solubility in various media (for example, in water or selected test solvents). In addition, these polyether chains (or more appropriately—these *polyoxyalkyl groups*) are integrally related to thermal stability properties.

Although the formulator's prime focus is, quite logically, the improvement of performance; a host of other factors (such as the possible side-effects which may occur under certain stringent formulating scenarios) must also be considered. One crucial factor, the possible loss of intercoat adhesion, is of paramount importance and will serve as the focal point of selected discussions in this Appendix.

It is common knowledge that, in certain application areas (including not only coatings, but also plastics and adhesives), selected silicone products can be effectively employed as "release media". Obviously, in such instances, silicones "adversely" affect intercoat adhesion in a planned, concerted fashion. The most prevalent release mechanism involves the *fixation* of silicone molecules within the top surface of a thin-layer system. In this manner, release papers and similar products can be easily manufactured on a reproducible basis.

How much control over intercoat adhesion is possible? In particular, how can one control adhesion with silicone additives? The answers to the above questions are within the framework of the following two-part overview of "reactive" and "non-reactive" silicones:

- "Reactive" silicones which chemically integrate themselves into the resin system can provide completely non-recoatable surfaces. Anti-graffiti coatings and release media are examples of candidate application areas.
- "Nonreactive" silicones which are incapable of reacting with the resin system exhibit the often desirable trait of migrating into the second layer; intercoat adhesion therefore remains virtually unaffected. (As a precondition of proper migration, the second layer must, of course, be a liquid during at least one phase of application and/or curing.) The exact mobility of the silicone depends upon both its molecular weight and its backbone/branching structure. Additional information about mobility trends and correlations will be provided in the second half of this appendix.

The Comparative Thermostability of Polyether-Modified Versus Polyester-Modified Polysiloxanes

Polyether groups based upon ethylene oxide or propylene oxide are not always oxidation-stable molecular entities; this is, of course, common knowledge for today's

Oxidation Mechanism

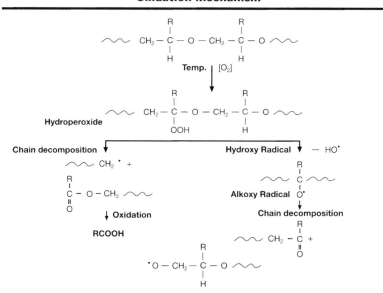

Figure V.1 Oxidation

modern synthesis chemist. For instance, certain polyethers which are exposed to temperature extremes in the presence of oxygen should preferably be stabilized with antioxidants. Traditional polyether oxidation chemistry is rather complex, and can be explained, in part, through a series of discrete chain decomposition steps in which specific radicals (as shown in Figure V.1) play important roles.

The decomposition process may furthermore result in the formation of reactive groups (–COOH) which, in turn, contribute to the formation of volatile, low molecular-weight by-products. Nearly all reactive groups which have been formed become free to interact with the reactive groups contained within the resin system itself. The final result is the irrevocable formation of an insoluble silicone layer upon the very top surface of the coating film. In other words, a previously "adhesion-unaffecting" and normally unreactive silicone product has been transformed (because of thermostability considerations) into a reactive product which now functions as a virtual "release layer".

What possible approaches might alleviate the aforementioned transformation? One very promising avenue of approach in answering this question lies in the realm of comparative structural chemistry. For example, one can examine various chemical subgroups and substituents for varying thermostability trends. Within the context of further pursuing this approach, polyether-modified and polyester-modified polysiloxanes have been studied at length.

Although several encyclopedic volumes could be written to fully explain the vast array of idiosyncrasies and variables associated with the thermal degradation of

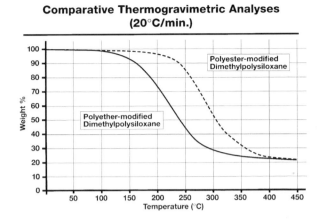

Comparative Thermogravimetric Analyses
(20°C/min.)

Figure V.2 TGA

siloxane derivatives—*thermogravimetric and infrared analyses (as shown in Figures V.2 to V.4) sufficiently subsume all major parameters into two descriptive variables— (1) weight loss and (2) the formation of IR-identifiable "COOH" and related chemical entities.* The following synopsis discussion of both variables begins by overviewing the Thermogravimetric analyses (TGA) and infrared studies which were employed.

Thermogravimetric Analyses (TGA)

The comparative thermogravimetric analyses (displayed in Figure V.2) demonstrate the vast differences in weight loss between polyether and polyester moieties. In regard to the *polyether product*, minuscule, yet easily detectable decomposition begins at 130 to 150°C; rapidly accelerating decomposition takes over at temperatures above 160°C. (Please note that minuscule degradation does not necessarily result in deleterious intercoat adhesion results.) In contrast, significant decomposition of the *polyester product* begins at approximately 230°C—one hundred degrees higher than with the polyether product!

Infrared-Analyses

The comparative infrared-analyses (displayed in Figures V.3 and V.4) demonstrate the vast differences in IR-curves of polyether versus polyester moieties. Four curves per base structure are displayed; each individual curve reflects the molecular structure present at a different temperature.

First of all, the *polyether product* was studied. As evident from the transformed temperature-dependent peaks—vast molecular changes occur as one progressively

IR-Spectroscopy of
Polyether-modified Polysiloxanes

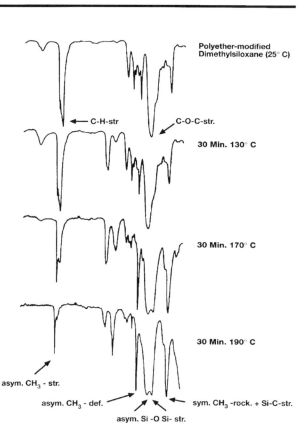

Figure V.3 IR study (polyethers)

moves from room temperature to 190°C. The strong "CH" peaks of the polyether groups disappear and the "CH₃" peaks of the "Si–(CH₃)₂" groups remain. At the same time, the rather interesting "Si–O–Si" peak, which is normally masked by the "C–O–C" peak, becomes more and more visible after the polyether groups begin to decompose. The most important change occurs in the 1600 to 1800 cm^{-1} range. Of particular interest is the formation of reactive, adhesion-affecting "COOH" groups. In the case of a typical polyether product, the resultant molecular structure after 30 min at 190°C matches quite closely the structure of an unmodified dimethylpolysiloxane! Of course, the aforementioned "decomposition product" most certainly results in deleterious intercoat adhesion properties.

In contrast, the *polyester product* demonstrates virtually no change as one progressively increases the temperature. No reactive groups are formed; in addition, silicone mobility and intercoat adhesion remain optimal.

IR-Spectroscopy of
Polyester-modified Polysiloxanes

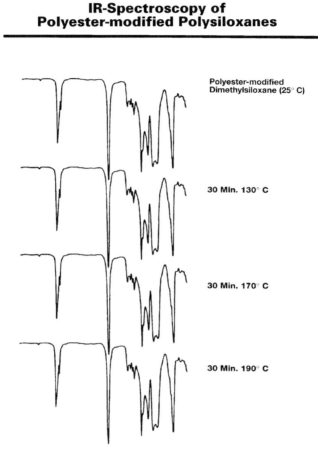

Polyester-modified
Dimethylsiloxane (25° C)

30 Min. 130° C

30 Min. 170° C

30 Min. 190° C

Figure V.4 IR study (polyesters)

Intercoat Adhesion Effects

In order to exemplify the dependence of intercoat adhesion upon the twin variables of temperature and molecular structure—a series of practical evaluations was performed. Metal panels were coated with a "silicone-containing" alkyd/melamine system. Several different silicone products were evaluated throughout the course of the test program. For the purposes of the following synopsis, only a limited number—three products, respectively designated as "X", "Y", or "Z"—of the tested silicones will be discussed. A control system (without silicone; designated as System "C") was also evaluated.

Silicones "X" and "Y" are polyether-modified polysiloxanes; silicone "Z" is a corresponding polyester analog. The concentrations of "X" and "Y" were held

System Features ("C" – "Z")

Control System "C" (Without Silicone)	
Surface Tension	30.3 mN/m
Silicone "X"	
Polyether-modified dimethylpolysiloxane	
Active Substance:	0.05%
Surface Tension:	27.9 mN/m
Silicone "Y"	
Polyether-modified dimethylpolysiloxane	
Active Substance:	0.05%
Surface Tension:	27.3 mN/m
Silicone "Z"	
Polyester-modified dimethylpolysiloxane	
Active substance:	0.10%
Surface Tension:	28.0 mN/m

Baking Conditions

First Layer	
Temperature:	130°C - 200°C
Time:	20, 30, 45 and 60 min.
Second Layer	
Temperature:	150°C
Time:	30 min.

Figure V.5 Integrated summary of system features and baking conditions

constant at 0.05% (weight percentage based upon the wet coating). In comparison, the concentration of "Z" (the polyester analog) had to be elevated to 0.10% to accomplish approximately the same surface tension reduction ($\sim$2.0 to 2.5 mN/m) achieved by the polyether products. All panels were processed in a gradient oven. (Please note that the gradient oven is a special instrument with multiple, microprocessor-controlled heat zones. A single panel, for instance, can be evaluated—during one oven-cycle—at multiple temperatures.) A broad battery of tests, especially in the realm of intercoat adhesion parameters, was performed; nevertheless, *before an adequate description of the test results can be presented, an integrated summary of System Features (for "C" to "Z") and of Baking Conditions is essential.* Figure V.5 provides the aforementioned summary.

As described, the first layer (a white coating) was applied on four different panels and then baked for 20, 30, 45, and 60 min. The second layer (a blue coating) was baked onto all panels at a constant temperature of 150°C for 30 min. The intercoat adhesion between layers one and two was evaluated not only with a classical "crosshatch tester", but also with a Single Projectile Launching Impact Tester (sometimes abbreviated as "SPLITT"). In order to gauge the intercoat adhesion, a projectile of

Reduction of Internal Adhesion/Damaged Zone (mm²) for Silicone "X"

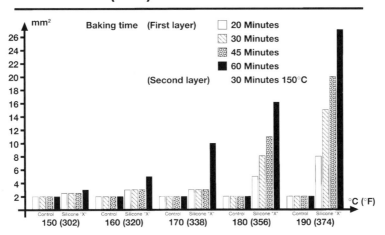

Figure V.6 System "X"

Reduction of Internal Adhesion/Damaged Zone (mm²) for Silicone "Y"

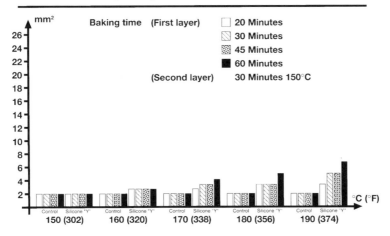

Figure V.7 System "Y"

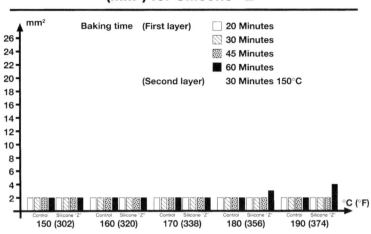

Figure V.8 System "Z"

defined mass (3 g), diameter (3 mm), temperature (25°C), velocity (100 km/h), and delivery angle (30°) was employed. Given the degree of test specificity, extremely reproducible results were obtained. Afterwards, the measured and tabulated surface areas of the "damaged panel zones" were compared (as shown in Figures V.6 to V.8).

The composite test results displayed in the impact-test diagrams clearly demonstrate the differences among silicones "X", "Y", and "Z". For instance, silicone "X" displayed early damage at temperatures as low as 150°C; subsequent damage rapidly increased commensurate with temperature elevation. In comparison, silicone "Y" fared somewhat better even though it was still a polyether product. (The reason for its improved performance was the presence of fewer branched structures.) Silicone "Z" (the polyester product) displayed—as expected—overall superior performance.

Conclusions and Practical Implications

The temperature-dependent oxidation of polyether-modified polysiloxanes results in the formation of reactive decomposition products which are derived, in part, from the polyether groups themselves. Such reactive moieties can intermesh with the resin, therefore adversely affecting intercoat adhesion properties. Of course, many practical implications can be derived from this fact. Multi-layer systems exposed to temperatures in excess of 150°C should not normally utilize polyether-based products unless the initial layers are designed to function as release coatings. In contrast, polyester-based products provide a wider range of formulating options in high temperature scenarios. Repair operations (even in situations where only single-layer systems were originally applied) can be particularly sensitive to intercoat adhesion variables*. In fact, some of the more bothersome "on-line" quality control disturbances may arise from some type of intercoat adhesion problem.

---

*There exist literally dozens of potential intercoat adhesion problems; the problem of siloxane degradation represents only one of many possible investigation areas. In fact, most adhesion problems arise from reasons totally unrelated to siloxane degradation. The reasons for loss of adhesion may include a host of factors ranging from inappropriate application techniques to improper substrate cleaning and preparation.

Index

Edward W. Orr is Senior Product Manager (Paint Additives) for BYK-Chemie USA. His background includes the synthesis, evaluation, and marketing of performance enhancers for the coating and allied industries. He is the author of over 115 publications and holds global patents for wetting and surface-control products. Mr. Orr was granted a B.S. in Chemistry from Radford University and an MBA from Virginia Polytechnic Institute.